Blessings

Prov 3:5-6

# BACKYARD SECRET

# EXPOSED

## A JOURNAL OF MY HEALING PATH BACK TO LIFE

## BETH STURDIVANT

ONENAMILLION, LLC

Published by
ONENAMILLION LLC
PO Box 455, Pineville, NC 28134 U.S.A.
info@backyardsecretexposed.com
www.backyardsecretexposed.com

Title ID: 4466088
ISBN-13: 978-0989840217

Library of Congress Control Number: 2013917121

*Printed in the United States of America*

This publication is designed to provide accurate and authoritative information with regard to the subject matter covered. It is sold with the understanding that the publisher is not engaged in rendering legal professional advice. If legal advice or medical expert assistance is required, the service of a competent professional person should be sought. (From a Declaration of Principles jointly adopted by a Committee of the American Bar Association and a Committee of American Medical Association and a Committee of Publishers and Associations)

Photos are reprinted with permission.
First round editing: CarolineDGreen.com
Cover Art and Design: Tracey Coon/noondaypublications.com
Book production: ElevitaMedia.com

This book is available at quantity discounts for bulk purchases. For information, call 704.458.1010.

This book is dedicated to:

My phenomenal children, Brandi and Brandon, who acknowledged the nature of my sickness, and aspire to live this lifestyle change with me as well as help educate the people around us.

My husband, Johnny, thank you for keeping our family afloat financially while not giving up trying to grasp all the drastic changes we endured.

My parents for unconditionally supporting me through all my endeavors.

I love you all!

# CONTENTS

## 2011

## 2012

## 2013

# INTRODUCTION

I was the middle child out of five who could do anything my mind was set to do. My parents used to say, "Beth is our 'can-do girl.'" I started early as an entrepreneur. In the ninth grade, I sold candy to my classmates. By the time I got to high school I was turning over $200.00 of candy a day—and I never ate my profit. I had two book bags: one for books and the other for fifty pounds of candy. I eventually earned enough to buy a new Nissan Sentra and pay for my own insurance too. You can read more about my school days as an entrepreneur in a section ahead.

Shortly after high school, I became the youngest bail bondsperson in the state to start her own company. A local TV station followed me around for eight hours and did a news feature on me taking a bond skip to jail. But after three years I needed more to do, so I opened the first of two Subway restaurants I owned and operated. A couple years later I remodeled a building and opened a salon with ten styling booths, three nail booths, a tanning room, and a massage room. With the bail bonding company, the Subway restaurants, and by then, a coffee shop, and two small children plus a husband at home, you could say I was the definition of high-energy. I was a multi-tasking perfectionist too. I painted my baseboards twice a year, vaccumed my furniture weekly, and kept all drawers neat and tidy. My husband once said he married a plumber, painter, and landscaper who could fix anything as well as prepare meals quickly without making a mess. When we moved into a beautiful new house—that changed our lives forever. I was very busy and endorphin-happy keeping my yard perfect and my house nice. From painting, plumbing, building a playhouse on stilts, changing carpet, you name it; I did it all. They called me "Project Girl"—I always had a project going.

But my attention and desire changed. I went from being all that I could be to not caring so much. I felt like I was going crazy in my body. I became very sensitive, angry, and annoyed. Then I couldn't remember things, like numbers, which was particularly unusual because that had always been my specialty. My daughter would say, "Mom what's going on with you?" Things got progressively worse until I couldn't even go to my store; it just made me feel anxious. I was starting to experience pregnancy symptoms and I would swell more while I was in the store—even my employees were getting a little concerned with how I was acting. Then the big day came when I literally felt my brain shutting down.

Looking back on my experience I'm amazed that I survived this nightmare—because there was no help available for me from anywhere. No one understood what was going on with me. I'd go to doctors and they couldn't tell me anything. I almost gave up.

Don't ever give up. If you're going through something you don't understand, you can survive it. I pray that my story will not only help you become aware of the invisible entity which almost took my life but that you, too, will come to realize that you can overcome forces greater than yourself.

# 2009

# FEBRUARY 2009
## Lonesome Highway

Blue lights appeared out of nowhere! I did not know where I was. Should I pull over or should I not, I wondered? "Thank you, Lord, there is a gas station." I needed gas anyway, so I got out to pump some gas. The policeman told me to get back in the car. I told him I needed gas. He said to wait and get back in the car. I was driving a nice Pewter H2 Hummer. I wondered what he wanted. "Ma'am, do you know why I am pulling you", he asked? "No". You were speeding, he replied. Well, I did not realize that, I said. I am kind of lost and can you tell me where I am? He said we were in Troy, NC. I wanted to know in case something happened. I am always cautious. Especially, since I was a bondswoman for 13 years. I am always thinking deceive and protect. I had been riding for at least an hour because the area looked familiar. He asked me if I had been drinking and I wondered why he was asking me that because it was only 11:30 PM. However, I was wearing an I.D. bracelet showing I was of legal age to drink. I said I had a birthday toast but nothing really to speak of. He said to wait and that he would be back. I was thinking, "What in the world is happening?" I had never been asked the question if I had been drinking. He said he smelled alcohol. I thought that was impossible. Well, he had me blow into the Breathalyzer. I blew .09. The legal limit is .08 in North Carolina. He told me I was under arrest and I needed to step out of the vehicle. I told him I needed to get gas in my vehicle. He said I would have to get it later. I knew the gas station would be closed when this episode was over, so I was thinking of how rude he was for not letting me get some gas in my car since I was already parked at the pump. Another officer pulled up and the first officer had him move my H2 to a parking spot.

I was escorted to the local jail, which was also the police station. I was dressed in my skinny jeans, wearing a nice top, with boots having 4-inch heels. I looked very nice. One officer even said, "Those are some good-looking boots." I said "Thank you." He asked if he needed to help me. I replied, "Absolutely not", you are not going to say I can't walk." Therefore, I sat down, unassisted, with two officers as they played on their Breathalyzer machine, trying to get it to act right. I will add that they never searched me or took anything from me. I was not fingerprinted, nor did I go into the back of the jail to a holding cell. I spoke with a magistrate who let me sign myself out on an unsecured bond.

I texted my daughter and told her I was in jail. She called immediately because she did not believe me since I still had my cell phone. So, I let her speak to the officer. I then called my husband, Johnny, but he had been out with some of his friends and they had been drinking, so that would never work in getting him to pick me up in Troy, NC. Well, I then called my girlfriend, Missy, and the official gave her directions to come and get me. I really did not know where I was, but it took Missy a very long time to get there. The officers left me in the waiting area of the jail; no one was anywhere around as I waited for my ride.

As I sat waiting, I was in shock. This really took me by surprise. I could not believe what was happening. They took my drivers' license for 30 days. Being stripped of my license made me realize I was charged with a DWI. My mind was just going crazy and I was trying to process the fact that I just got arrested for a DWI. I just wanted to scream! Here I had worked all day and didn't even really feel like attending that birthday party. I drove at least 45 miles to get to my old classmate's 40$^{th}$ birthday celebration. I had one drink. While I was at the party I felt like my skin was crawling and I was feeling anxious. I felt I had to get going because I was feeling crazy and overwhelmed in my body at the restaurant bar. When I left the party in Concord headed back to my home in Fort Mill, I was on Highway 27, which is called Albemarle Road in Charlotte. As I continued driving, I kept seeing Albemarle Road street signs, which I was familiar with since I grew up in the Charlotte area. My mind raced thinking thoughts of what was going on with me. I had two Subway restaurants to run and now I wasn't supposed to drive since my license was suspended for 30 days. In addition, there was my son who was getting his permit in 18 days. Suddenly, I thought, "Oh great! how am I supposed to get my car back home since Missy was coming to take me home! How was I going to pull this off?" When Missy gets here I thought to myself, I will drive her truck so she can drive my H2 Hummer in case the police are waiting to see if I am going to drive. I did hear them say they would be getting off in an hour. Understand, I was 70 miles from home and it was 2 AM.

Missy finally arrived and I was so grateful for her coming to get me. She said she would drive me back in the morning. I said no because I have to open my restaurant in the morning. I did not have time for that. So we decided to go and see what the place looked like, and if there were any police officers around. I said to Missy, "You will get in my car and I will just climb over into your driver seat without getting out of your truck so I can follow you, and we need to get gas somewhere close. I am about to run out of gas."

We were in a little town and there was nothing open in Troy, North Carolina. The gas station where my Hummer was parked was closed, just like I thought. Finally, we came across a gas station that was open. Well, I looked down at the gauge and saw that Missy needed gas too, so I had to put gas in both vehicles. We then drove to Missy's house because it was closer than mine. Plus, it was early and I thought that maybe I could sleep a few hours before I had to open the Subway.

I was so nervous, I was beside myself. My mind was racing, as I felt really crazy in my body. I had difficulty understanding what I had been feeling, let alone be able to describe the sensations I was experiencing. Four hours went by and the sun was rising, so I told Missy I had to leave to open the Subway restaurant. I would wear the shoes and uniform that was in the Subway office since I didn't have time to go home. I was a little nervous about driving with no license. I really did not have a choice because of my responsibilities as the owner and manager. I was feeling beat up and just overwhelmed; wishing I could rewind the previous night.

I got to my Subway restaurant and started setting up the front line with all the vegetables, condiments and meats. I started the bread-proofing machine to bake the bread for the day. As I started to prep food, my phone rang. It was my dad. He asked, "Have you heard from your brother?" No, why? My dad said, "I just got a call from a jail. It was collect, but I could not understand them." I thought, what in the world, this is freaky. What a coincidence! Thinking to myself, does he know about me? There is no way. It had just happened and I had not talked to anyone. This was crazy. My dad had never called like this before in the early morning. My brother had never been locked up before and why would anyone be calling my dad collect anyway? It was, I was sure, a fluke. However, still a crazy coincidence considering the circumstances. Dad said, " Alright, I am going back to bed. Have a good day." I continued with my opening routine to have everything ready for the day. I opened up at 9:00 AM. My first employee did not show up, neither did he call. My second employee didn't show up nor call as well. There were only a few customers so that part was okay but I wondered what was going on and if I were being punished already. I had always been able to handle working by myself because this is who I am.

Jeffrey, my third employee, showed up for work. He said, "Where is everybody?" I said that I did not know. Jeff told me that it was supposed to snow that day. I said, What? Are you serious? "Yes, Ma'am", he replied. I had not seen the news so I did not know the weather prediction for that day. So I thought this is why the employees did not show up or call. That realization made me feel better. I was thinking about not having a driver's license. I was going to have to be extra careful driving later that night in the ice/snow. We started to have a little lunch rush, but nothing that Jeff and I could not handle. The day went on and around 5 o'clock I told Jeff that it was looking pretty bad. I knew he had a longer distance to drive than I did. We needed to start closing up

for the night and get home. We finished up and I had Jeff to follow me. He had to drive past my house anyway to get to his.  I
didn't want my employees to know about my DWI until I got my
license back. You know how kids and adults talk. I did not need
any trouble and I still needed to drive. I thought I would tell them
about my night out once I got my license back. I am a "keep it
real" boss and I wanted them to learn from my experience. But
we would have to wait 30 days for the truth to come forth. For
me, my employees are like having many children. I feel that I have
to help mold and prepare them for their adult life as well as for
future employment.  Just like parenting, we are a reflection of
how our children are taught and I felt that being a first boss to
many, holds the same responsibility.

I had one month to drive without getting caught. Having been
a bondswoman and owning my own business, Marie's Bonding
Company, for thirteen years, I had gotten many people out of jail
for DWI charges. In my lifetime, I never dreamed I would be
faced with the same charge. I was about to experience what it feels
like to dodge the police while driving the roads. I had to drive to
get to work every day. Unfortunately, heavy construction near us
brought speed traps and checkpoints. I got home after having
been gone for almost twenty-four hours. I will never forget this
experience.  I just wanted to go to bed and wake up from the
nightmare I had just experienced. When I woke up, I realized this
wasn't a dream and I was going to face it head on as required. I
needed to take a deep breath and understand this was not the end
of the world. Many things could have happened. I decided to hold
my head up high and keep on trucking. I could not believe that
when I blew the Breathalyzer, it registered a .09 with just one

drink. This was very disturbing and made no sense to me. The Subway was closed due to the snow and I needed to get some things done.

I continued to drive to work. Furthermore, I noticed that I was swelling awkwardly in my body. This was very weird and I needed to focus on my health instead of the DWI. I knew my husband could take our son, Brandon, to get his driving permit just like I did for our daughter, Brandi. I was going to work, being very careful about driving. I took the longest route with no short-cuts and tried not to drive unless absolutely necessary.

That weekend, Brandon had a basketball tournament out of the city. My husband was going out of town and our daughter was away at college. I took him, along with some of his friends. While we were at the game; they kept winning, so we broke for lunch. On the way back to the gym, we saw what looked like a license check up ahead. Here I had these boys in the car, Brandon's friends and teammates. They had no idea what we were feeling seeing those blue lights. As Brandon and I looked back at each other, we saw that it wasn't a license check, but there were a bunch of police cars surrounding a vehicle on the side of the road. We both looked at each other. Brandon said, "Mom, my heart had fallen to my feet." Yes, me too, I said. You know it's one thing if you just drink and drive with no life responsibilities. I, on the other hand, have a lot of responsibilities and have always had a busy life. This was very nerve-wracking to endure and I swore to myself that I would not be in this position again. Thankfully, we finished the tournament and arrived home safely.

On the following Monday, Brandon went and got his permit with his daddy. Later that night, we had to go to his school and get something to eat, so I let him drive. We were on our way back home, going down this country road, and the blue lights came on while Brandon was driving. Great, what is it? We pulled into a church parking lot and the officer told us we had a tag light out. He asked for our license and registration. I immediately got out of the car and asked him to please not scare him, as this was his first drive. We laughed, but I was trying to distract him so he would not ask for my license since it was suspended. Everything turned out okay and we both let out a sigh of relief. My nerves were shot and I was ready to get home putting all this behind me. I did tell my son no matter what, he should never drink and drive. Maybe this was a lesson for his future.

# MARCH 2009
## Holding On

I went back to Dr. Scott Makerson, my OBGYN doctor to get another exam. My previous ultrasound showed I had a cyst on my right ovary. Therefore, he wanted to do another ultrasound to see if the cyst had shrunk. In the meantime, I had been going twice a week to Dr. Zhou, a Chinese medical reflexologist. I was seeking help for the cyst and for dark spots that were everywhere on my skin, especially the great big one that was in the center of my forehead. I was also experiencing lots of unexplainable swelling in my body. My mother went with me to this doctor's appointment to view my ultrasound. They saw that the cyst had not grown, but

had shrunk a tiny bit. We went into the conference room to have
a discussion about the cyst and my weight gain. I was really upset
about gaining weight. I was ready to have surgery if he thought it
would help me lose weight. The doctor said it would not affect my
weight. He said that I seemed more concerned about the weight
gain than the cyst. I said, "Yes, I am not a fat person. I have been
extremely weight-conscious all my life." In my brain, gaining
weight made me feel like a big failure, worthless, and wanting to
commit suicide. Gaining weight provoked a real deep emotion
that I have been tortured with my entire life. He recommended
that I go to see a nutritionist. This made me furious and ready to
end my life. I said I had already gained 20 pounds in six weeks.
Something was wrong and it was not the food. I now felt that this
doctor was not seeing the full picture. He did not pass my test at
all and was not listening to me. He said he was going to order
blood-work and that he wanted me to go to the Blumenthal
Cancer Center downtown at Carolina Medical. He would have his
secretary make me an appointment. He had never seen a cyst that
looked like mine and he wanted to get a second opinion. I agreed
and scheduled an appointment for the following month. I did
notice when they took my blood that it was extremely dark. I told
my mom this was a waste of my time and money. I knew nothing
more that day than when I went in for my first visit. I cried all the
way home, telling my mom I was done with life. I did not feel
good at all. I was swelling by the second and I just wanted to die.
I was feeling bad for saying these statements to my mom. I knew
this was making her feel helpless and hopeless. My mom told me
I could not end my life, I was going to get through this and everything would be okay. I listened. I was silently talking to myself as
positive as I could, considering how sick I was at the time. I must

say that the human mind is very powerful and can harm you when you really start to build on anxiety, fear, and pain. I could not understand why I was feeling so sick and anxious.

I tried to take my mind off my pain and swelling as I prepared to go with my dad a few days later to get my license back. I continued to work, though feeling very sick, with constant nausea, nervousness, and severe headaches. It also seemed as if I would swell more when I was at work, but felt even sicker when I was at home.

My dad went with me to Troy, North Carolina to pick up my license because my 30 days were up, and boy was I thankful. Dad drove as I wanted to use my new Dell laptop since I had a wireless card for Internet access. I noticed that I felt kind of sick and my stomach felt hard as I was using it while it laid on my lap. I said to my dad that it seemed like the computer was making me feel sick. He said to put it up. I always work at using my time wisely; working on whatever has to be done so I can be productive, making the most of every opportunity. We finally arrived at the courthouse Clerk of Court office. I paid my fine and they gave me back my license, scheduling my court date for the month of May.

As we left to drive back home, my dad asked how in the world I had gotten all the way out there. I said I didn't know but the signs stated it was Albemarle Road. Dad, being a former truck driver, said if I would have kept going, I would have ended up in New York! I am not good with directions. I can follow written directions, but not East, West, North, and South or numerical signs. Dad drove an eighteen-wheeler. That's his specialty, not mine.

When we got back to the Charlotte area, I dropped my dad off and went on about my day to see how the Subway was doing. I was feeling nauseated all the time. I had all the symptoms of pregnancy, including my breast leaking, but the doctor said I wasn't pregnant. My son told me he was glad I wasn't pregnant. He didn't want a brother or another sister. I said, "If I'm not pregnant, then something is really wrong." "Like what, Mom," he asked? "I don't know," I replied, "but I'm gaining weight by the minute and I feel sick all the time.

# APRIL 2009
# Dell or Bust

Another week went by and my friend Omega went with me to the beach for a girls' weekend getaway. I drove, of course. I had my license back and I always drove. I started to feel sick, so I said to Omega that I wished I knew what was going on with me. I did not feel right, I was gaining weight, and I had all the symptoms of pregnancy. Furthermore, the doctor said it had nothing to do with the cyst. We got to the beach and checked into our room. We went to the grocery store to get some snack foods. This was to be a kick-back-and-relax weekend, watch some movies, and enjoy some good old girl talk. I brought along my Dell laptop computer to work on my Ancestry tree. I loved doing this and I had already accumulated a thousand people in my tree. I am always doing a project, as project is my middle name. I started working on the laptop and mentioned to Omega that every time I was working on the computer, I felt kind of sick. Suddenly, I made the connection and thought, "Bingo!" The light bulb went off in my mind. We looked at each other and both said at the same time, "This is probably what caused the cyst." I stayed on my computer and about 45 minutes later I felt really sick. My stomach was hardening and swelling. I looked like I was nine months pregnant. My friend, Omega, said "Let me put this up." So, I waited about thirty minutes and started feeling better. I wanted to try again to be sure what we were seeing was actually happening. I got the computer from the bathroom in our hotel room. About fifteen minutes later, I started to feel sick; my stomach swelled up and got hard again. Omega jumped up, took the computer away, and told me that I didn't need to get back on it again. I agreed with her. We called my husband and I told him that we had figured out why I was feeling sick, nauseous, and why my stomach was swelling. We told him everything and he believed us. However, I wondered if anyone else thought we were crazy or questioned if this was what was really happening. I told him I loved him and would talk to him the next day. Omega and I were still in a little bit of shock after witnessing what had happened. I was telling her that when I Skype, Brandi says I look really pregnant.

I did feel sick and nauseated, with a headache. I kept having a crazy nerve sensation running through my whole body. The dark spots on my skin had gotten worse. I did not know what was wrong. We drove back home and Omega said she would take my computer with her and keep it in the closet. I could get it another day. She offered to research the swelling issue and see what was on the Internet. I agreed telling her I would stay off the computer, but to let me know what she found out. We had a great time together, as always, but in my mind I was worried. This was not normal for me to be feeling sick. I needed to get home since we were closing on our house the next day with the local power company.

Tuesday arrived. My husband and I went to meet the Energy Company representatives at the law firm, who had contracted to purchase our house. In return, we would rent the home from the power company for the next three years in order for our son to be able to stay at his high school. They had inquired about purchasing our house back in October of 2008 on the understanding that they were going to add more power lines. They needed more right away. I was excited to sell so that we could pay off some debt. However, I was also sad because over the last nine years I had done a lot of landscaping design work in our yard and all the decorating of the house. My children had grown up in that house and we had a lot of great memories with pool parties and numerous family gatherings. We closed on the house, which took about two hours, including the driving time. I was feeling really sick and not my usual self. It was the nausea and nerve sensations that were the worst, along with headaches that would not stop. By now, I was starting to stay out of work and run the restaurant from my phone. This was all I could do, because I felt terribly sick.

When I woke up the next day, after closing on the house, Omega called the house phone because I had turned off my cell phone. She told me I would not believe what she had found out and preceded to tell me that the heat from a laptop is a form of radiation. This made sense to me since the cyst on my ovary looked like it had been zapped in my ultrasound pictures. Omega also found some disturbing information about houses that are too close to High Tension, High Voltage, and Transmission Power Lines. The symptoms that some people experience includes headaches, nausea, nervous feelings, body feeling hot, sleeplessness, anxiety, depression, dizziness, reproductive system issues, and the list continued. I said, "Wait a minute, are you telling me that I am sick from these power lines and that I have a cyst from this computer?" Yes ma'am, she said. Is this why the power company purchased our house, I asked?

They knew there were big chances of a problem, but we had no idea. What was I going to do? We had just agreed to rent for three years. I could not even process the fact that I was in a poisonous house. I had raised my kids there. Could this be what was making me sick and the reason why my daughter was having so much trouble with her menstrual cycles? I was in shock and in a state of disbelief. I was also angry to think that the power company was acting like they didn't know what was going on with my house. I told Omega that I cannot use my cell phone, or be on the computer. I would have to figure out what to do. She said it looked like a lot of government officials and scientists were denying the effects. I thanked her for looking this up and told her that I would talk to her later. My friend Missy called telling me she had been researching online about the computer and asking if I knew that the power lines were hurting me and that my neighbors were getting hit too.

I am getting sicker now just thinking about these issues I face. I really felt trapped and sadly disappointed that we had even bought the house. How could they build a house so close to these lines and say it was safe? The county and city had both issued permits. I was thinking that the county and the city must not have known the effects of the power lines. I was going to give them the benefit of the doubt in order to ease my mind. I told her to continue to do the research, but I needed to go lie down. I was getting really sick by then, both mentally and physically. My mind was racing with thoughts. I was thinking so fast, trying to piece this information together. My mind was going back nine years. I was trying to process all this information and fix the issues at the same time. This was like something you had no time to prepare for, like a sudden death.

Spring break for high school kids was approaching and it was now time to take my son Brandon to the beach for his spring break. I contemplated how I was going to do this trip with his friends? I felt very sick and was also frustrated about my situation. I had been taking Brandon and his friends to the beach for many years. I enjoyed being a part of his life and appreciated that the boys were close to me. I did not want to disappoint them or to be the cause of them missing out on some fun during their break. So, I needed to go to the Subway first, to check everything out, to get change for the register, and to complete the paperwork before we left. I got it done. Although it just seemed like I was slowing down. I felt sicker, with more nausea, nerve sensation, and a headache. I finished doing what needed to be done at Subway with Brandon helping me. I began to notice as soon as I walked in the restaurant that being in there was really making me sick.

I took Brandon home to pack and wait for his friends. I left and went to my mom's. When I got there I sat down at her kitchen table feeling as if my life was changing in front of my eyes. It literally felt like my brain was moving and processing information in slow motion. I had never experienced this before, and I was having trouble with my handwriting coordination. This was really scary and I didn't want to say anything to anyone about it in case it was going to pass.

I asked my mom to call this person that my brother, Barnabas, knew to see if the magnetic resonating stimulation pad he had would help me. The representative answered stating he would be there within the hour. He came over, and after trying it out, I agreed to rent the pad. He also rented me a Gauss Meter so I could see the electrical and magnetic fields of things around me as I felt the nerve sensation throughout my body. I lay on the pad for eight minutes. The rep showed me how to operate and pack the pad. Brandon and I were going on our yearly spring-break vacation to the beach. My friend Missy, along with her son Zach, Brandon, his friends TJ, and Kevin all rode to the beach with me. While driving, we could not have the radio or any cell phones on as these seemed to make me feel worse.

When my head would hurt, I had my friend turn the Gauss Meter on to see what it was reading. I then realized a scary thing. There we were, passing and sitting at stoplights, passing power lines and Wi-Fi towers, as well as hearing some loud noises. As all of this was unfolding, my body was reacting and it was teaching me about exposure to a world that few people recognized. This was frightening to discover. Once we arrived at the beach, I rented two rooms, one for the boys, and one for Missy and me. I was thinking I could get on that body pad for eight minutes every four hours while trying to relax. I thought I would have enough time to get better while the boys ran around having fun for the week. Missy and I did go to Wal-Mart late that night because I assumed there would be less people and the store would have more lights turned off during the late hours of the night. I was really feeling sick when we got done shopping for food and was ready to get back to the hotel room so I could lie down. Fortunately, the boys came, unloaded the groceries, and we went on to bed. I was getting a little rest, and as always, I maintained a feeling of being positive no matter what happened. I was thinking this couldn't be that bad and I would get through it like everything else in my past. I would be better soon. It was just a bump in the road. I continued to lie on the pad every four hours. Thankfully, the pad did ease my head from hurting a little. My daughter, Brandi came down to visit since we are about two hours from her school. Both of my children were watching me but did not say anything. They knew something was not right, but no one, including myself, really knew what was going on with my body. Finally it was time to pack up and head back home from the spring vacation trip. I did feel a little more rested and a tad better. I had hoped that I was getting better. We drove back home and I got everyone to their destination.

As Brandon and I got home and while walking into the house, I noticed that I could feel more, that I was physically in tune and aware more so now than before we left to go on the spring vacation trip. We had only been gone for one week. My helpful son unloaded the car putting most of the stuff away so I could call check in on the Subway restaurant and lie down since I started to feel worse. I called my husband, Johnny to let him know we were back and to tell him I noticed I was feeling worse since I walked into the house. I could tell he was bothered by this problem and thought I was crazy. He didn't directly say so, but I could feel his negative energy or vibe. When he got home, I shared with him what I had experienced and that I didn't feel good. It was time for bed but I couldn't sleep. It felt like my body was on fire. I was beginning to put this all together. Johnny told me I felt like a furnace, but I did not feel like I was hot or running a temperature, just my body was hot to the touch. I was constantly getting up because I couldn't sleep. I had not slept well for about a year by this time, and I was just now realizing why. Those power lines had been interfering with my sleep. My neck ached for years. All those new pillow purchases and going to the chiropractor for adjustments did not seem to help relieve the pain altogether.

It's difficult to explain how sick and crazy I was feeling. This scared me because I did not know what reactions to be on the lookout for, nor was I aware of anyone who had experienced this type of illness. In order to find some form of relief I decided to go down to my husband's sports room, or man cave, to test the electrical currents on the Gauss Meter. I found that it was the lowest electrical currents room in the house and the furthermost distance from the power lines. Therefore, I lay on the pad there, and tried to sleep. My head was in excruciating pain, I was really hurting. Well, it was time for work, but I couldn't go. I had to run the store by using the house phone to call the employees and get some more help in the store. My free labor was now being replaced, which increased my restaurants' payroll. Unfortunately,

I was not able to use my cell phone because being on it made me feel worse. My husband did not understand at all and was giving me a silent negative attitude, which made me feel even sicker. I was in an abnormal health crisis, unable to physically work, and could not go to the Subway restaurant.

I began to think of others and felt sad for all those who suffered with illnesses and still had job duties to perform. This really takes character and inner strength to keep going when you do not feel well enough to work. Regardless of how I felt, I did have to get dressed and go to our restaurant in order to program the register with the new prices launching for the five-dollar footlongs. This also required that all the menu boards be changed. Being home sick that week, I asked my husband to bring me the menu boards for both Subway restaurants so I could get that done for our businesses.

My husband had never programmed registers or changed out the menu boards before, so I got all the prices from our Subway headquarters and wrote them on a sheet of paper in order to make this process move more efficiently considering my difficult health crisis. This was torture having to tend to such minor details and I wished he knew how to do those little things instead of me being the only one with the knowledge.

I was feeling like I had thousands of needles poking me all at one time. My headaches were really bad, and by this time, I was becoming sensitive to lights and hating noises. My body was doing this upchucking. I felt as if I was losing my mind, time was flying by, and I just wanted to cry.

Well, I managed to get a game plan together in order to go program the registers. I had my mom, Brandon, and his friend, TJ, go along to help me. Of course Brandon drove, which was a nice relief. He was always there when I needed him. I was sick and he was a brand-new driver. A Hummer H2 is not a small car to keep in the lane while driving down narrow country roads. I was a little more relieved knowing he had lots of former practice with riding lawnmowers and driving four-wheelers. So, I gave everyone their jobs before we got there and tried to explain the reasons I was acting like this, which was really out of character for me. I did not know much myself at this time either, but I figured out that I was reacting to everything electrical. I explained to TJ that I was having trouble with electricity and that I could not stay in the store for a long period of time. When I said it was time to go, I needed them to be on point ready to walk out the door. I got in there and started to do the register. I had the sheet of paper with all the prices written out so I could move quickly and not have to go between computer screens searching for the new prices of the subs. I tried to stay focused, getting done what I came to do, but I was feeling sicker by the minute. Now everything electrical was bothering me. I lasted about 45 minutes then it was time to leave. Brandon drove and my mom told him to go a different way to take her back home. I started to do that upchuck thing over and over, almost like convulsions all the way to my mom's house which was about a 20 minute drive from one of our restaurants. I could just feel the concern in the car from my mom. My son was just being as strong as a young man could possibly be. I felt like a beaten down wet noodle. When we finally got to my mother's house I got some water, laid down a while and rested while the boys played basketball. I could not explain what was happening. I was so worn out, depleted of energy and unsure why I was having this horrible reaction. I just felt really sick and majorly confused. I was very worried and felt bad that the two boys had to witness my upchucking reactions. It sounded like I was throwing up and I hated to hear that myself. They were good boys who were always looking out for me. We left my mom's, dropped TJ off and went home. I was miserable and scared. Brandon told me that everything would be fine. Such a positive little hero I had raised. My husband did not know what to think. Was I crazy or was this really true? I went to bed and had another bad night. I could not sleep at all. I thought about what was going to happen the next day and wondered what I was going to learn.

My day began and I continued to feel miserable as my head felt very tight and extremely uncomfortable. I could not think clearly. It felt as if I was in a mental fog. Everything I did felt as if it were slow motion. I was sitting on this pad and writing notes. I felt like my body and brain were shutting down. I could not get on the computer and thought I may never be able to do so ever again, which was very difficult for me to conduct my business and personal affairs without the convenient use of technology. I was using my Gauss Meter, looking at the time on the clock and began to notice a series of patterns with feeling more pressure in my body according to different time periods. It appeared to me that when I started feeling more pressure in my head and body, the meter readings went up in digits. I believe the power company increased the need for more electricity in the area through the power lines out back of my house at different times of the day and night. I was beginning to piece this part together because I had that sicker than usual feeling about 5 a.m. every morning, and at lunchtime as well as dinnertime. I had pressure in my body, the headaches were worse, I was painfully nauseated, and I also had the upchuck thing going on continuously.

I was sitting there, trying to stay calm and put mind over the matter. I was talking to myself and saying that I needed to go to Mom and Dad's to get some relief. I was fighting in my mind. I needed to do this, but I would be walking away from my teenage son, husband, and all of my belongings. I continued to lie on the pad every four hours, recording all the Gauss Meter readings, and feeling terrible. I did this for a few days and got worse with nerve pain, headaches, and severe upchucking. Literally, it would seem like a few minutes, then the whole day would be gone. My husband, Johnny, would be back from working all day and I felt like he had just left for the day.

# MAY 2009
# My Father's House

I ended up packing a bag and going to my mother's house. As I was leaving, I was upset about the fact that I was upchucking and could not stop. I had to drive on a two-lane country road and could not pull over. This is when I realized that I had nowhere to run or hide since electricity is everywhere. This was one of the most frightening experiences of my life. The ride seemed longer than usual and I felt I was shutting down more so than before with my cognitivism. I was missing my turns and wondering where I was going, and then remembering that I had taken this same route for the past 9 years.

I was in extreme pain, compounded by the stress of not having a choice in leaving my husband and children in order to escape the high levels of electricity in my own home. I arrived at my parents' house and went to lie down in their downstairs bedroom. I tried to calm myself down from this mess that I really did not know anything about my situation other than the fact those power lines behind my house were very dangerous. To top all this, I was constantly swelling by this time, with no hope in sight that I would stop. It was out of control.

I was still running our Wesley Chapel Subway from the phone. My mom collected the money and brought me the closing receipts. My husband also continued to bring me all the receipts from both stores as well as all our mail from four different mailboxes. I started to be aware of my speech slurring, and my head hurt worse than if I had suffered with a migraine. I couldn't deal with lights at all. I did not like noise and could not watch TV as it made me swell. I would shake when I tried to write, and it would take me about eight hours to write a check. I would just sit, stare, and look out the window.

My mom continued to try to help me and brought my food to the bedroom. I couldn't even walk into the kitchen by this time without swelling and getting pinged in the head. We all started to see that every appliance, including the night-light, was giving off an electrical field. I felt like a missile was going through my head every time the refrigerator went on and off. My breasts, upper arms, hands, legs, ankles and feet, along with my whole mid-section kept swelling. I felt like I was going to burst like a balloon. What was I going to do? What was this, I kept asking myself? My daughter arrived back home from college for summer break. It had been about three weeks since I had seen her at the beach. My son would spend the night with me so I could drive him to school. I was sick, but he was my son and I needed to do what I could for him. Who knew how long I would be here? I will say it was only unconditional love, because driving was getting real difficult. I didn't know how to deal with whatever was happening to me because my nerves were getting more sensitive. I continued to swell and do that upchuck thing. My daughter, Brandi, saw what I was going through and told me that I could not drive or be on my phone at all. She said that she would drive Brandon to school for me.

Brandi took my phone and started to answer all my calls, telling the callers I could not be bothered. In addition, Brandi decided not to attend summer school for the next three months, and would lie on the bed beside me just to hang out. She was trying to take on the role of being me as much as she could in handling my affairs. In the meantime, she was researching online trying to find some answers on what to do or how to help. Unfortunately, Johnny was being standoffish, not saying much, but would call and flip out at any given moment when something went awry at one of the restaurants. I could see the concern on everyone's face and the lack of understanding my suffering. Brandi found this drink concoction from the United Kingdom and its purpose was to treat radiation poisoning. There was also a recipe for making a detox bath. The drink recipe consisted of one teaspoon each of baking soda, cream of tartar, and sea salt mixed with 8 ounces of hot water. We would let it mix and do this explosion thing. According to the instructions, I could have this drink every two hours. I took my first taste of this drink and I thought it was great because it reminded me of a soup broth. I charted every time I had a drink. My body actually started craving this mixture.

I then started the detox baths twice a day. The AM bath was 1/2 cup of bleach and 4 cups of baking soda in water as hot as I could stand. The PM bath consisted of 2 cups of rock salt and 4 cups of baking soda, again as hot as I could stand. I did this procedure and lost from two to four pounds while I was in the bath, due to my profuse sweating. I did this drink and bath for weeks. After about two weeks, I stopped with the bleach, thinking it was too toxic. I continued the other items and began to include the Himalayan salt detox bath in accordance with the full moon. This was thirty dollars a bath, but I wanted to get better and was willing to absorb the expense. I continued these solutions for the next few months. I did experience an unusual and frightening situation one time while I was taking one of my baths. My mom decided to vacuum the downstairs while I was away upstairs taking my bath. When she did, I started to feel this electrical sensation and it got worse. I just started crying uncontrollably and my body felt like I was getting electrocuted. I was yelling for help, but no one could hear me. Finally, my daughter came in to check on me and wondered what in the world was going on with me crying. She immediately ran downstairs and asked my mom to stop vacuuming. She came back and saw how much that helped me. I was feeling very nervous and weak. It took me a minute to get it together so I could even get out of the tub. This was crazy. I could not believe I could feel the vacuum cleaner. I was super sensitive to everything.

Over the summer college break, Brandi decided she wanted a dog that she could train and take back to school with her to enjoy. So, she got a little Yorkshire terrier, like my dog BJ, and named him Polo. We kept him in a child size playpen so he did not urinate on the floors when she had to leave for a while. I was too sick to even babysit a little dog. I know she did this so she did not have to focus on me. It was a happy distraction for her.

My Aunt Barbara, from Minnesota, and Cousin Rhonda, from Nebraska, finally arrived at my mom's after having a hard time with the directions as they had driven a very long distance. They made it to the house after enduring hours of traffic and wrong roads. I had seen my aunt and cousin in October 2008, which was seven months earlier. My aunt looked at me in shock saying she could not believe how I looked and that I was so swollen. She said that we must figure out what was going on with me. What they saw was a changed skin tone, a very swollen abdomen, cognitive issues, and my slurred speech. I couldn't remember what I was saying or comprehend what others were saying. This was really bad and we were all clueless. I continued to lie on the bed in pain feeling sick and did an upchucking thing at any given moment without notice. It was like my body was trying to throw up but nothing came up or out. This was such a horrible and terrifying ordeal to experience let alone not fully understand what was happening.

This was such an over the top ordeal that my mom's oven and range top was causing me to upchuck when it was on heating up food. In order for her to cook, she had to shut my bedroom door, the kitchen and dining room doors while opening the front and back doors just so she could make a barrier between the kitchen and me. She also opened the door in the breakfast area and the door between the kitchen and the garage, to let the electrical waves outside. This kept the electric field smog from being trapped in the house. We later realized she had to turn the oven off at the breaker box when not in use because I couldn't even walk into the kitchen without going into an upchucking episode. The coffeemaker couldn't be on or I would swell. I could not walk outside, as the sun made me swell something fierce. I was wearing sunglasses all day and at night because I could not deal with lights. Lights, when turned on made my head just pound. I could not sit on the living room floor, as I would start swelling because the well pump under the house was electrical and the TV was in the same room. Talk about one of the most absurd and unbelievable ordeals, this is what I experienced. I was kind of like an electrical pyramid. I noticed I couldn't sit on one of her couches because my head would hurt so badly from the nightlight that was in the bathroom behind the wall where the couch set. The electrical wires in the wall bothered me. This was only one of my many observations. We were detecting these things with the Gauss meter. So every time I started feeling sicker, we would take the Gauss Meter out and get the electrical reading. I also learned that I reacted horribly to my mom's three-way lamps and the dimmer switches throughout the house. I could not walk into her kitchen without feeling as if I had been zapped in the head by the metal strip for the sliding doors. This was really making me crazy. I could see that the doorbell was also causing me to have a reaction. Now who would have ever thought that a doorbell could cause electrical problems? My dad was constantly running around trying to see what was bothering me and shutting many things off in their breaker box. My dad had to disconnect the electrical garage controllers that were across on the other side of the house. My parents live in a nice size house out in the country but that didn't matter. I was learning that electricity was everywhere and that I could not hide. My brother, Barnabas, came over to build my mom a fence around her garden. While there, he used an electric saw that he plugged in across the house on the back deck about 150 feet from where I was in the bedroom. I could feel myself swelling and I started to upchuck. I was getting worried but I knew I had to stay strong. I was familiar with the mind over matter philosophy and worked it every day of my life. This all went on for weeks.

Then weeks turned into months. Now my court date from the DWI was coming up and there was no way I would be able to go in my condition. I had to get the date postponed, so I called a law firm asking what I should do. I was informed that if I got a doctor's note and paid $150.00, they would represent me. I was unsure what doctor I could get since I did not know one except Dr. Meehan, who is a chiropractor and has some knowledge about EMF sensitivities. I decided I would get him to write a letter in order to let the courts know I was physically ill and needed an extension. So my mom called him and thankfully he agreed to write the letter. My court date was extended until the end of June and I thought I would have time to get better not really knowing how severe my situation was or how long it would last.

# JUNE 2009
## The Book

My son completed the ninth grade. Now our family was somewhat lost since no one knew what to expect with me. I was still running the restaurants because my husband was avoiding the facts that something was terribly wrong with his wife. He believed that if he put off accepting the fact that I was sick, it would not be happening. I was getting frustrated with him because I couldn't do anything and he was worrying me even more by being in denial. When he would call me, he would make negative comments. This was not the time for any negative words especially in my condition. He hardly spent any time with me. I knew he didn't want to hang out for any length of time at my mom and dad's house. However, that was where I had to stay and the children spent most of their time there as well. BJ, my Yorkshire terrier, was also affected by the situation. He was losing his fur, breaking out in rashes, and acting very nervous. He would not leave my side and obviously knew something was not right with me. I was watching him experience the same nervous problem that I had, which was the inability to deal with loud noises at all. BJ would run and hide his head under the blankets or furniture. Mom and dad were both worried, feeling helpless and hopeless about my situation. My restless leg syndrome started kicking in and boy did that hurt. It was also very annoying. I asked my dad to rub my legs because they hurt so badly. I could never find any lasting relief. He did rub my legs and it helped as long as hard pressure was applied like a deep massage. The pain was uncontrollable, which drove me crazy. I decided not to go back to the oncologist at the Blumenthal Cancer Center. I felt like I would die if I had an MRI;

therefore, I called and canceled my appointment. My daughter called and said she had a dream that I had died. She told me not to go to that appointment. I told her I felt the same way and had already canceled. Now that was total confirmation I had made the right decision not to go.

The doctor, during the first appointment, told me I should keep an eye on my cyst. As long as the cyst was going down in size and my blood work looked good, I should be okay. Brandi was reading more about how electricity and radiation affects everyone. In addition, she found that when a person becomes sick and sensitive this is a major game changer. During this time, I continued to receive treatments from Dr. Zhou, a Chinese medical reflexologist. He said, "Beth, your body is really going crazy. What is going on? Your endocrine system, thyroid, thymus, pituitary, and nervous system are all falling apart. You have bad headaches, insomnia, weight gain, and the list goes on and I know I need to fix you."

My family and I had been going to Dr. Zhou for four years, since 2005. Therefore, he knew us. He had helped me with my back and shoulder pain when I fell off a four-wheeler. I had hurt for two years prior to him fixing the pain in just one visit. This is just an amazing way to help the body through your feet. This time it did not seem to improve me. I knew something was really wrong with my body because I was gaining weight eating healthy as I always had for years. Brandi ordered a book for me on EMF sensitivities. I could hardly wait to get some literature to help me better understand my situation. I could not use the computer. A friend of my parents, Joe, came over one day. He is a retired pilot,

so he understood a little of what was going on with me, certainly
better than the rest of us. Joe told me that I needed to order
Diodes for the cell phone, computer, breaker box, and a copper
Pyramid Diode that would help protect the entire house. These
diodes would allow me to use my cell phone that had Internet
access so that I could do research on my situation. In the meantime, I would continue to work at staying calm and focus on healing.

My Aunt Barbara noticed that she was being shocked a little
from the bed we were on and I noticed that I was being poked.
What was this, I wondered? It turned out that Mom had a magnetic mattress on the bed. My dad took it off and stored it in their
recreational room. I did feel some relief when this came off the
bed, but it didn't make a huge difference. I needed a task I could
do in slow motion without jarring my head and one that involved
no thinking. I decided to sort my mom's pictures that have been
stored in boxes for over 40 years. I would go to the rec room and
sort for a couple of hours at a time during the daylight since I had
a tough time with lights. This would help to get my mind off my
problem. I love to do projects and stay busy. Well, Mom had 45
years of pictures in those boxes. It was kind of rewarding to go
back in time and see relatives that I did not know. I continued on
this job for a while. I would swell a little so my time was limited
each time I sorted. I did finish this task and waited for Mom to
decide what she wanted done with the different groups of photos.

My Aunt Carolyn, Mom's sister, was coming down from
Minnesota with her three grandchildren. We had a house full of

people with my Dad's sister and niece, me, my two children, and our dog. Now, my mom's sister and her three grandchildren would be staying too. Mom does enjoy company but it was a lot of negative energy in the house with someone trying to heal. I visited mostly with my dad's sister and cousin. My Aunt Carolyn did not understand what was going on with me and was busy with the grandchildren. She didn't stay long as they were leaving to visit my brother, Ben, and his girlfriend, Nan, in Charleston. Now I had many people telling me what to do, or what I needed to take for my situation. I had to learn to listen and not take it the wrong way. I appreciated the fact that everyone cared enough to suggest something, anything to help me. Also, my family members did not need to take my behavior personally because I was really sick with severe cognitive issues.

Finally, my book arrived and I was excited to get it: The Electrical Sensitivity Handbook by Lucinda Grant. I started to read it and the first thing it talked about was how electrical fields (EMF) are making people sick. My first question was answered - power lines made me sick. Now what should I do? The book went on to give a lot of science and facts that were way beyond my understanding. There were not enough details to say what to do about it if you were experiencing certain problems. (Page 49) "ES is not yet a medically recognized illness in many ways and therefore traditional health/disability insurance payments, legal avenues, and federal/state benefit programs are difficult to access". Electrical Sensitivity (ES) needs to be included under the umbrella of environmental illnesses for Social Security Disability and other benefits for future public policy. (Page 50) "In 1994, the American Bar Association Journal reported that EMF cases were being filed at the rate of one per month. These were primarily power line EMF cases. Is there a cure for ES? If you find one, please let us all know. As ES becomes better known and accepted, benefits will eventually follow. There are ways to improve your health and reduce symptoms but there is not one clear path for everyone besides avoiding those things that bother you". (Page 51) "If the home is unlivable/ unhealthy for the environmentally ill person, then the patient feels they have no safe place to be away from EMF exposures. This is very stressful and frustrating. Sorting out which factors are bothering you is very helpful on the road toward feeling better again". (Page 52) "Storing up electricity in the body can cause stress". (Page 57) "(The body is electrics magnetic.) When a person is ES, their adaptability to other EMFS is impaired. The use of EMF's treatment of illness by conventional medicine is in the early discovery stages". (Page 58) "The trend is called early medical or energy medicine". (Page 62) "Children living near high voltage power lines show increases in leukemia". (Page 70) "Doing daily living tasks then become very difficult or impossible without help". (Page 71) "Fluorescent light is a common irritant to ES patients". This book can in no way describe the broken dreams, failed marriages, lost careers, lost homes, shattered family relationships, ridicule, and hopelessness that haunt the environmentally ill. Wow, what a roller coaster ride to read about and actually feeling this trying to avoid these negative facts. I made the decision to rearrange and overcome this negative challenge into something positive in my mind because this is how I am.

I cried to myself, trying to be strong. I was too young for this and my children needed a mom, not a burden. I could only imagine what would happen in my marriage. What was I going to do? I am crying even as I write this thinking about all that I read and how my life will be impacted. This is a nightmare for anyone to go through. I had no idea what was in store for me, how to figure this out, or even where to start. All I could imagine was how unfortunate this was for us all. I lay there for a while, just thinking about everything. My husband came by; brought our paperwork, and lay down on the bed beside me. I told him what I had just read; especially the part, "Is there a cure for ES? If you find one, please let us all know". He lay quietly, thinking and doing his normal "Umm, umm", then pausing. He told me everything was going to be okay, with tears in his eyes, as he hugged me. I could not explain all the emotions going through my body. I was feeling just devastated and wondering if my life was over. I did not want to be a burden on anyone. I just wanted to go to sleep and never wake up. He left and I continued to ponder my life while staring out the window like an Alzheimer's patient. I shook as if I had Parkinson's disease and definitely had cognitive issues both illnesses emulate. I kept reassuring myself that my children needed me to get it together. I immediately started to think about how I could make this negative situation into something positive. There had to be a cure. My mind, attitude, and beliefs have always been positive. I strived to think and speak the opposite of what was taking place at the moment. There were flashes of ideas running through my brain.

So, the journey to my healing began as I was remembering that if God made our bodies to try to heal itself until our last breath, then this was no different. I needed to figure out the right ingredients to turn this around. I immediately started to understand that I had a big job to accomplish. I envisioned myself sitting on the beach with my daughter saying, "Wow, were those five years something?" I saw my finished work. Now I just had to get there. I started to read the book again and highlighted what I thought was important. I tried to digest as much information as possible. I was having a very hard time with reading comprehension. I could read, but could not remember enough to even repeat what I read to anyone. I knew what I had read, but was unable to communicate it back in words. I just knew that I would need to start taking notes, pictures, and videos. In my mind, I did not want to be video recorded. I was so swollen and crazy looking, but I knew I would never relive this moment so I needed proof of what I was experiencing. I needed to find some doctors who understood EMF and could prove my medical condition was indeed ES. I would need to keep data of what my symptoms were and what I did to stop them. I was turning this into a project. One day, I would tell the world that there is a cure for ES. I would bring it to the world. I feel God uses people in their own ways. My parents brought us up to understand the body and the importance of sunshine, food, exercise, fresh air, sleep, and to respect how the body works. I knew that if someone was sick, they needed a lot of rest and a minimum of stress. I continued to get my thoughts together and to change my helplessness into hope. I prayed and thanked the Lord for allowing me to be that special person. He entrusted me with this task.

# JULY 2009
# Hope Pad

My husband took over the operation of the restaurants. He finally understood and accepted that I was not really myself. He is the type of person who avoids fearful issues by not talking about them. He doesn't like to discuss serious illnesses or the death of someone close to him because of because he is afraid to face reality and deal with how these issues may affect his life. Therefore, he blocks it all out in order to cope. The majority of his coping skills, just like the rest of us, come from our upbringing. Children do live what they learn. Good or bad habits, we are all products of our environment to some degree. As my husband took over the restaurants, he noticed that the monetary change fund was coming up short. When my mom and daughter conducted the money count each day, it was hard for me to tell what was short. I really could not be worried because I was too sick. Happy Fourth of July, I thought. My husband counted the money and it was short again. He had only been doing this for two days. The employee who was responsible thought that my husband would not figure this out and that we were not going to catch him. After working for us six years, this employee was very self assured and did not believe we would ever suspect him. I then had to go and pull the cameras up and search for a thief. After closing, this thief would let himself into the store with another ex-employee and use a knife to try to wedge the money out of a slit in the top of the safe. I watched him help himself for about 30 minutes. I printed the picture so he could see that we were on to him and would have solid proof in case he denied it. My husband called and told him that he could either pay us back or go to jail. He paid the money back and lost his job.

This shocked me to the core because nothing people do had really surprised me in the past. I would have to say this was sick. I had been a great boss, always there to help the employees through mentoring; modeling each of them great work ethics as well as teaching job skills that they could take and use anywhere they went in life. I was hurt and my husband was furious. He partially blamed me, which caused me unnecessary stress. I felt as if he should have been there monitoring the employees and restaurant more, then the theft would not have happened. Either way, I had to focus on healing. I believed that everyone would be repaid for their wrongdoing in due time, and I knew that I couldn't fix anyone's emotions that caused them to do hurtful things to others, including me.

The diodes arrived. As I took the Pyramid gadget out of the box, I felt relief. As I put the necklaces on, plugged the gadget into the outlet, and a diode on the breaker box, I felt relief. Placing the diodes on my cell phone and computer, also brought relief. Note: do not get the word relief twisted. These items were helping me cope for minutes at a time. In my mind, I was excited to be able to research for five minutes. Those minutes would eventually add up. The more information I could gather, the quicker I could come up with a remedy. Now I could start gathering research and dive into this journey of healing. I was still taking care of both restaurants' paperwork and checkbooks, as well as being a mom to my children all from my new bedroom at my parent's house. As I began to research, I soon discovered that EMF dangers and ES sicknesses were being ignored and disputed by government officials. Apparently, based on certain research information, our government uses a term called "Prudent Avoidance (PA)" which created a set of safety standards to keep this comprehensive subject at bay (www.officeoftechnologyassessment.com). In addition, power companies were denying that there were any negative effects known on the human body. The United States government had set safety standards, which they were following. However, the standards have not been updated to consider all the new 21st century electronic gadgets that were having more potent effects on the nervous system and cells. It is very hard to read what is being denied when you are physically going through the pain, suffering, and isolation. I quickly understood how this information could create an epidemic of fear. Furthermore, there was no way the government or the electric companies were equipped to handle such a problem. We all require electricity for our everyday needs. I am a logical thinker. I began questioning why I was the only one being affected at this magnitude and the other people in my home were not experiencing the same things. Therefore, what was going on?  What was so different about me? I came across some information about Amalgam fillings and how they can affect the brain and body as a conductor to the brain itself.

Mercury is very toxic and I needed to have my mercury fillings removed. At this point, I was concerned because my teeth were hurting. All of my teeth felt like they were going to fall out. I told my daughter that I was going to lose my teeth and she would say, "You are not". I would then say, "You do not understand how bad my teeth hurt, they really feel like they are going to fall out". I continued to read and study but I could not go to the dentist right then to have them removed. I was very sick and super

sensitive to electricity. I would just have to wait. While researching, I came across information about how all twelve of our merid-
ians go through our teeth. I was looking at the tooth chart to see
how they were positioned, such as the ears, hands, feet, eyes, and
spine. All parts of the body have pressure points known as alarm
points. When we experience pain, it will show up in one of the
alarm points. This was absolutely amazing to me to see that God
had designed us with such complexity. I thoroughly enjoyed any
kind of science that I could put together that would address my
condition. Removing my mercury fillings was a top priority considering the toxic release they give off, but I was too sick to handle having my fillings removed at that time. My mouth had a
metallic taste, my breath was annoying, and I was in constant pain.
Weeks had gone by and it was now the end of July. The court
date had come up again. I went, and it was extended for another
month. I will say that I was about 70 miles one-way from the
courthouse. This was the most expensive birthday toast I had ever
had. I could hardly ride in a car without going through those
upchucking episodes along with having massive headaches, and
feeling very nervous. What a horrible experience that was to
endure. I had not ridden in a car and I now knew why because it
made me really sick.

By now, I had known Dr. Michael Meehan, our family chiropractor for about 12 years. He was the one who my brother,
Barnabas, got the information from for a natural magnetic resonating stimulation pad. This is the same pad I had used months
earlier on Spring Break vacation with my son and his friends. I
went to visit Dr. Meehan and he became more interested in this
pad once he saw it helped me some relief from the constant pain:

I went there to have him examine, document, and videotape me about EMF/ES symptoms in order to have some kind of professional doctor's opinion. Furthermore, I chose Dr. Meehan because he was the doctor for this type of pad on the east coast. Dr. Meehan had studied health issues related to EMF/ES sicknesses; therefore, his knowledge was helpful. While I was in his office, my stomach was swelling. I was grateful to see his acknowledgement of the swelling and happy to get it on video. I understood many people were not familiar with EMF/ES sickness and would assume I was crazy unless I had documented proof. Thankfully, he recorded the video of me making mention of what was taking place before our eyes. Since he knew I was coming, he had shut off the X-ray machine along with all of the lights. Even with all that done, I still swelled. Dr. Meehan's office is located in a fairly large business park in a three-story building. I appreciated his efforts seeing how super sensitive I had become to surrounding electricity. I had gained about 30 pounds by this time, which depressed me more because there was nothing I knew or could do about it at that time. My children had to purchase some larger clothes for me to sleep in because I had outgrown my entire wardrobe in a matter of weeks. So, after 22 years of being the same size, one can understand how disturbing this was to me. Especially being a woman who has worked at keeping her figure and health.

# AUGUST 2009
## Special Gift

Well, I had to move from my mom and dad's since I kept getting worse. We tried all we knew to do in order for me to get some form of relief but nothing really resolved the situation. I couldn't drive down the road to my house without upchucking. In August, we celebrated Brandi's 19th birthday at my parents' friend's Chinese restaurant, Wan Fu. I had not been to a restaurant in a while. They agreed to put us in a private, back room and turn off the lights in that room in order for me to be able to come. I always did put mind over matter especially when it came to my children and certain memorial events in their lives that I wanted to experience with them. As a parent, I loved being with my children. Just having them as my special gifts every day, brings me so much happiness. Therefore, missing my daughter's birthday wasn't an option. Although, I had to be careful and go through some extra pain, the memories could not be recreated once they were done.

My friend, Caroline, came and noticed that I was not my normal self. I had not seen her since recovering from my back going out, which was about seven months earlier. We were talking and catching up. I told her about my situation and that my parents' house was not helpful in my recovery. She offered for me to come to her house if I could tolerate her electricity. We did know that she had a much older house and the breaker box was a lot smaller. I prayed I could tolerate it because I was really suffering in pain immensely every day.

# SEPTEMBER 2009
## Head Sap

After going to Caroline's house, I did discover I could sleep on her couch in one position. I also was able to sit up on the right side end of the couch. However, I could not sleep in any other room of the house because I could feel myself reacting to the electricity. I began to understand a little more of what was going on and how to prevent extra pain. She was good at keeping all the lights off and not running the kitchen appliances while I was there. We ate a lot of salads and cold food.

My brother, Barnabas, introduced me to a water ionizer machine that increased the alkalinity of drinking water. We had a sales representative come over to tell us about the machine and what it could do. I asked a few questions about the business side as well. Why not, I thought, I'm sick and I need to make money. Caroline and I understood that with my purchase under her, she would be able to get a water ionizer machine for her home with the commissions earned from my order. However, her machine would go on a layaway payment option until enough commissions were accumulated. This was a great option for both of us because when we started to drink this water, I noticed that my cognitive thinking started clearing a little. I was drinking between two and three gallons a day. My body was craving this water and I was able to stop the other alkaline drink mix that I was consuming. I did have another company's brand of water ionizer machine, but it did not deliver the same results. They are not all created the same is what I came to discover. So, by now, two weeks had gone by and I started to have this sticky substance similar to honey or ear wax leaking from my head every day, all day long. This was very unusual, so Caroline and I videotaped what was happening. This was another incident that others may not have believed, thinking I was insane if I had not captured it on video. Fortunately, it started to ease up some and only leak every few weeks. When this happened, I would get a really bad headache the day prior. This headache was awful and like no other before, so I saw a pattern now. I knew my head would have this sticky substance and the headache would stop; although I still maintained a constant headache, just not as harsh. As usual, I was still managing the checkbooks for the two restaurants and trying to be a mom from my friend's house. I would have emotional breakdowns periodically. I would cry; have a pity party with myself, then get it together. I would always feel relief when I shed some tears, not knowing then that it was chemicals being released.

Weeks went by as I continued to live with this electric magnetic sickness. I continued to research, and went to court twice as they postponed my DWI case each time I went to court. Caroline and I spent many hours on the couch. I was helping her as best as I could heal from her reaction to coming off all prescription medication that had resulted in her gaining over 100 pounds in less than a year. We always joked with each other, saying we were like two old sick people in a nursing home. Even though our situations were not funny, humor made them easier to deal with since we were both fairly young. I had to keep myself busy by continuing my quest for knowledge through research and writing numerous notes.

My son was now in 10th grade and on the football team. It

was opening night for his school in their brand-new stadium. I did go to the game and was able to stay for about twenty minutes sitting on metal bleachers. Most of the parents had not seen me in about six months. I looked sick. My friend, Barbara, said, "You look like someone on a lot of medicine, grey, and puffy." I could not deal with the stadium lights and all the loud noise at all, which triggered my upchucking episode. I left and went home very sad. I was thinking about how these were Brandon's last few years of high school. He had been playing football since he was eight years old, and now I would have to miss watching him play. I was very upset about that, along with the fact that I was not living with him in our own home. I didn't get to see him during the week because school, sports, and homework were keeping him very occupied. This was very emotionally difficult to endure since I am my children's mom and was accustomed to always being with them.

My brother, Barak, came to Caroline's house to learn about this water ionizer machine. He said "Oh my, if you say you look better, you need to be careful because you look really bad." I thought that was not a nice thing to say to someone sick. I knew I may not have looked better, but I did feel a little better. He always would speak his mind, regardless of how it would be perceived by the other person hearing the words. I refused to let his statement get me down. Everyone knew that the weight gain was the hardest thing for me to deal with emotionally while going through this health crisis.

# OCTOBER 2009
## Happy Birthday to Me!

Time went on and my 39$^{th}$ birthday arrived. My husband came by to take me to dinner. We went to a restaurant that was close by where I was staying. I tried to find a table and spot that I could deal with, but it did not really work. Johnny got irritated with me because I was not dealing with the electricity and WIFI going on around me. With all the cell phones in the restaurant, and the power lines outside on the street it was unbearable for me. He refused to come to grips with my situation. He lived in denial in order to avoid facing the fact that his wife was sick. I was hurt, angry, and frustrated with his lack of support and refusal to want to understand my situation, especially after six months of this going on with me. I told him to just take me back to Caroline's and to go on about his business. I couldn't help or control what was happening in my body. I was very upset, sad, and just sat in the dark finishing out my birthday evening alone, deeply disappointed. I knew that I was up against his coping skills, which were emotionally tough to manage. However, I chose to be thankful that I was still here for my children. Getting upset was not good for me, or my health.

# NOVEMBER 2009
## In All Things Give Thanks

The holidays came and I decided to go back to my house to my son and husband. Caroline's mom came back home after being gone tending to her sister who was ill during the whole time I had been there. In addition, the house was small and too crowded for all of us to live there together. Therefore, I went back to my house and purchased gadgets to be able to cope with my illness not knowing that the house was really too poisonous for us to be there. The EM fields were terribly high; however, I used this rope that helped in shielding the bed where I slept along with outlet and breaker box gadgets to add more protection. I also had the copper pyramid faraday, which helped with the vortex up to 100 feet in each direction. Obviously, no one could run the vacuum cleaner, use the washing machine or dryer, nor run the dishwasher and oven while I was in the house. We had house phones with cords and no Wi-Fi. I was forced to stay on my bed; unable to watch much TV due to the noise, vibration, and light. My eyes were still sensitive to the lights. I did this day in and day out for months. Most days, I would sleep the day away. I was so sick feeling that hours felt like minutes. I would often wonder if I were ever going to find real help. I would always reverse that thought and tell myself that I was getting better and gaining more knowledge. We all went to my parent's house for Thanksgiving dinner. Thankfully, my mother enjoys preparing meals and is a great cook.

# DECEMBER 2009
## Christmas Prayer

I was too sick to put up a Christmas tree and all the decorations. However, my son refused not to have one, and surprised me by doing it himself. He was a real trooper, always asking me if I needed anything. My daughter was still in college, but came home for the Christmas holiday break. She spent a lot of time lying on the bed with me, and continued researching trying to find remedies. My husband tried to avoid the reality of this big black cloud hanging over us, and I really had to work at ignoring him. It did bother me, but I do best not expressing myself when I'm irritated. I would sometimes just cry and pray while playing this wind-up music box that my mom bought for me years earlier. The song was "What A Friend We Have In Jesus", and the music gave me some peace and joy. I would also read my bible for peace, but that was very limited due to the light making me more nauseous. The verse that I quoted many times to get me through a moment was Proverbs 3: 5-6, "Trust in the Lord with all thine heart; and lean not unto thine own understanding. In all thy ways acknowledge him, and He shall direct thy paths." This had so much meaning to me, and I gave the burden over to my Lord and Savior. I really tried to focus on the Lord directing my path first and me listening. I had not been doing this very much prior to that time. This was a true wake-up call for me about having unconditional support. No matter what, I always found comfort with prayer.

We all went to my parents for Christmas dinner and gathering. We had to eat with no lights on because of my sensitivity. However, I had to stay away from their house until mom finished cooking. She tried to air out the kitchen from using the oven and stove for cooking. I knew that with the appliances being on in the kitchen that it would cause me to upchuck. I was learning to avoid certain situations and conditions in order to maneuver around the triggers that caused me to go into those upchucking episodes. I had a lot to be thankful for because my immediate family was totally supportive of me, and did not think I was mentally unstable. My husband had a little doubt in the back of his mind, which helped him to avoid the reality of what was happening. I understood because this was a sickness that had not been heavily discussed in our society. The "powers that be" claimed it did not exist. I knew from all the research I had read that scientists were denying any known cause of EMF sicknesses. My husband had no evidence or documentation that society understood or recognized the issues. However, I knew it was real, because I was the one suffering through all the mental and physical pain involved. People who knew me saw that the facts could not be denied about EMF sickness. I always kept in mind that I was raised to think outside the box, as pertaining to the body. On the other hand, my husband, friends, and extended family did not understand, but it was okay because they eventually would come to it by watching me. We had a good time at my parent's house as we always did. I was really quiet and not my usual self, but I enjoyed just being with my family.

# 2010

# JANUARY 2010
## Let It Snow

The holidays were over and school was back in session, so I was alone a lot. I continued to research and seek attorneys at law firms who could represent my case against the power company. I did not discuss the matter of the case with my neighbors and pretty much stayed to myself. For one thing, I had to stay on my bed trying to protect myself within my world of gadgets used to help lower the effects of EM fields that were bombarding my house and body. Unfortunately, our neighbors had developed negative attitudes towards us when they discovered we sold our house to the power company. The power company was buying property from all who were on the right-of-way line. Our neighbors were concerned about the market value in selling their own homes. Our house happened to be closest to that right-of-way line. I knew they did not know what we chose to do because none of them were as sick as I had become from the exposure to the power lines that were directly in my back yard. In addition, the home on one side of our house and further away from the power lines tower had a new family move in since the home had just recently been built. We had already been there for eight years before that house was built and were the closest to the power lines tower. I did feel sorry for all these neighbors' children who would eventually suffer from being in close proximity to the lines. To make matters worse, the elementary school they all attended had a Wi-Fi tower on its property. These children's bodies and nervous systems were not getting a break from the high doses of radiation exposure that comes from these modern day society necessities. I did not like to think about all these people who were affected by power lines and

WI-FI towers not realizing the health dangers associated from exposure.

I continued to use my cell phone as an Internet connection to manage my emails as well as do more research on EMF and ES, which was more tolerable because I had diodes on the back of the phone. I could then print documents to read when I felt better. However, I still could not do this for long periods of time because of my other obligations of managing our checkbooks for the restaurants and trying to be some kind of a mom to my son who was still in high school. I really could not do much housework because of the high levels of electricity and I was forced to live in my bed where most of the home gadgets were located to help me exist in this poisonous house. I was also missing out on Brandon's sports activities and his games. This was emotionally hard for me since I enjoyed being there for him, but I had no control over the situation.

It was now the end of January and I was fed up living in this house the way I was forced to live. I decided I would go stay with my younger brother, Barnabas. We had always gotten along well. He had tried to find help for me with many different things over the years. We refer to him as Dr. Barnabas because he has a kind and caring spirit, and has several "healing gadgets." He was at a loss for me also, but we were all trying to figure this out. Barnabas was very conscious of what bothered me and we both maneuvered throughout the house in the dark. Fortunately, I was able to cook on his stove because it was an older model that had no electronic or digital panel. I was still bothered by it some but it's hard not to do any activities at all. He would work all day and I liked having a meal prepared for him when he got home. My brother made fresh vegetable and fruit juices for me every day, as I could not handle the machine. It snowed one night and we went for a walk under the moonlight. The outdoors seemed really quiet with the snow on the ground. Unfortunately, I was not able to sleep well at his house because the heating unit really bothered me. I would wake up with my neck hurting and have a horrible headache. His unit kept running and I knew something was wrong. I eventually left and went back to my house. I really want to thank my brother for always trying to help me; I love him for all he has done. I felt that if I were uncomfortable, I should at least be at home with my son. Periodically, I would spend the night at Caroline's and she would go to court with me on that DWI case that seemed to carry on forever. This month we drove up there again and the case was postponed once again until

# MARCH 2010
## Legal Consult

Spring was slowly approaching and my son was going to turn 16. A year had gone by and I was not better physically. However, I had gained more knowledge about my sickness. In my quest of searching out my legal options, I got a referral for an attorney who was in Charleston, South Carolina where my oldest brother, Ben, lived. I called Ben and told him all about my EMF sickness and about going down to see this attorney. Ben and I had not seen each other in at least two years. He had no clue as to how sick I really looked. He only knew what our mother had shared with him. Ben spoke with someone who found another lawyer; Mark with the Peper Law Firm, who was interested in reviewing my situation. Thankfully, my brother scheduled an appointment for me, so I drove down to Charleston and stayed with him for a few days. Ben did not say anything about how I looked. He welcomed me and we hung out doing nothing but talking, just catching up on things. Well, we decided to take a stroll on the beach so I rode with him and as he was driving I had an upchuck episode. My episode intensified as we approached the main road that had so many power lines. I was under a huge attack within my body and head. Anyway, as we walked out onto the sand, a distance of about 15 feet, I started upchucking tremendously. Ben was calm and stood there with me, rubbing my shoulders telling me it was okay. He and I both witnessed my feet literally turning a burgundy color. That was shocking to see. My episode stopped after about 12 times and Ben suggested that we walk a little bit further in order to get out of the same spot of energy I was releasing. We believed that based on my research, the beach and ocean are known to help in grounding the body by pulling electricity out; therefore, this explained what was happening to me. Afterwards, we went back to his house and Nan was there. This was the first time I had met Ben's girlfriend and I did not like for us to meet under these circumstances, but it just happened that way. She works as a registered nurse at the Veteran's hospital in Charleston.

The next day we met with the lawyers, Mark and Dyllan. I told them as much as I knew about the research, which was not much, but I did have a lot of information about the physical pain EMF sickness caused in people exposed to the radiation. They decided they would take my case, but Dyllan wanted to first study the materials in order to further know how to proceed. Well, I was grateful but did not feel totally confident that they could do this case. For one thing, they were a new law firm, so funds may be an issue. Secondly, going against a power company would be a real challenge. I left their office feeling a little hopeful but not totally confident. As Dyllan studied the information, he also inquired with other law firms who would want to team up with us in order to share the financial burden as well as add their own knowledge and experience. Therefore, I ended up spending a lot of time with my brother throughout this leg of my journey until July. As the heat got worse, I could not bear the weather any longer. I learned how weather changes affected me. When it stormed, I would upchuck profusely and my head ached horribly. When we ate, we would go eat at the Early Bird Diner on Savannah Highway. I could only sit at one booth in the entire place in order to keep from getting too sick to eat. "Who deals with this," I kept thinking, as I tried to do the simple things in life.

# APRIL 2010
## Dentist Visit

At my lawyer's suggestion, I decided to apply for disability. I thought I would not qualify since my husband and I owned a restaurant. My husband was employed and I was part of the franchise. I called in April of 2010 and did a phone application. They told me what they needed to collect from doctors and that I needed to sign a release form in order to move forward. I applied and about six weeks later I received a letter stating I was denied assistance. I was upset in light of all I had suffered and continued to suffer with not being able to work producing an income. So, I called the number on the denial letter and spoke with a lady in that department. I said, "You do not understand, I cannot work! I cannot even deal with any electricity without getting sick. I can't even vacuum my house, nor do laundry. So how can I work?" She kept saying, "Ma'am, I am sorry. You can dispute the claim and we will resubmit it." She told me that I could start the process again and I would be required to see the doctors that they scheduled me to visit. I had no problem with that requirement.

I started going to another doctor, Dr. Kanelos, with Carolina Family Health Care. He was a friend of Dr. Meehan. He also had an office near one of my Subway restaurants. I knew him as a customer who frequently visited another business I had owned, my coffee shop in Weddington, North Carolina as well as my Subway restaurant. He asked me what I would like to accomplish. I told him that I wanted help losing weight and figure out this EMF sickness. He put me through a normal doctor's routine for first patients. He prescribed Lasix for my fluid retention and Clonazepam for nerves and anxiety. I got both prescriptions filled and started to take the medication. After two days, I started to ache in my hips like a throbbing pain. I talked to a friend about the aches and she said that the medication was pulling fluid off me. Furthermore, she said that if the fluid was not surface fluid, the medication would dehydrate my muscles, including my heart.

I stopped taking the medicine right away and the aching stopped a few days later. In addition, the Clonazepam was not soothing my nerves so I stopped taking that prescription too. I was not raised to take medicine so I really understood the side effects, but I was desperate and in a lot of pain, which is why I gave it a try. I continued to go back and forth to Charleston, having to sleep on my brother's couch, facing only one way. It seemed that in every house I went to, there was only one spot that I could tolerate and nowhere else would work.

In the meantime, I still wanted my mercury fillings removed because they seemed to be further aggravating my situation and illness. After researching for about eleven months, I knew mercury fillings were very toxic to me. I finally called and scheduled a dental appointment cleaning. By this time, I had not had my regular six-month dental cleaning done in two years. I had Nan, an RN, go with me and videotape my visit. I needed this as documented proof of all that went on when I am exposed to electricity. Furthermore, I did not know how my body would react being in the dentist office. I had sunglasses on, a circuit breaker diode on my forehead and diode pads down my body. I was covered in diodes. I was in a lot of pain, suffering with a headache, and nauseous. However, I had begun to build a tolerance to my pain and learning to manage it as well as could be expected. I have always put my mind over a matter when I really needed to do something. Dr. Phillippi did his exam and said that my teeth looked good. I told him I needed my amalgam (mercury) fillings taken out and asked if he could recommend a dentist. He said that he could do the job. I asked what his procedure was and what precautionary steps he took to ensure that I did not swallow mercury or inhale the vapors. I told him that when I called the office, I was told that they didn't do this procedure. He said that he did not advertise the removal procedure but he would take care of his patients especially if it was a medical concern. I scheduled an appointment for May and was relieved to know he could do this procedure for me since I had known him for the past 20 years.

# MAY 2010
# Mercury Rising

I got a call and a letter to schedule an appointment with an assigned medical doctor and psychologist for my disability case process. I went to the psychologist, whose office was located in Waxhaw, near my parent's house. He was very dry with no personality at all. He asked a series of questions not related to my health issues such as: Did I like myself? Did I know where I was? Did I know the date? Could I take care of myself? How old was I? How many children did I have? Where did I work? When did I leave work? Was I married? He held a penny in his hand and asked what he was holding. He held an apple in his hands and asked what he was holding? Were these questions typical standard questions, I wondered. He did not ask me anything about my situation even when I was having an upchucking episode in his presence. I did ask him if we could turn his lights off for me. Who knows what he was thinking while I was in my retching episode. My mom came with me for support and to make sure I would be OK. I had a bad impression of his report that he would be submitting to my caseworker. I walked out of there thinking that it had been a total waste of time. The next day, my mother and I went to the assigned medical doctor located in downtown Charlotte. He did the standard test that any doctor would do for a first visit. He was from India. I thought this would be better since it seems that people outside of the USA are more likely to understand things that harm the body. He met with my mom and I and he was very nice. He asked me to explain what was going on with my health. I told him that my house was too close to the high-tension power lines, which caused me to be sick. I was having trouble with all electricity. I could not work or deal with any items that had electrical power. I had to wear the sunglasses all of the time because light bothered my eyes. I felt like a thousand needles were poking my entire nervous system. I had severe headaches, nausea; my breasts were leaking similar to a nursing mother; extreme weight gain and dizzy spells when I felt as if my head was spinning. I liked working, but I was too sick and just did not feel good. He said he had heard of this but it sounded like I was more educated on the matter than he was. "Great!" I said, "So what are you going to write in your report?" He told me that he would say I had EMF sickness and would report these facts that I was sick. I said that would be okay and thanked him. He wished me luck and said he hoped I would feel better. I did not have high hopes with this visit either. Well, I had tried and it wasn't fair that I would be denied because of lack of knowledge in the medical profession. I would just have to wait and see what transpired. In the meantime, my son had a sports injury. We went to the doctor to see what was happening with his knee. The doctor told us that his ACL and meniscus were torn. He needed surgery. I really felt terrible for him. I do believe he was injured as a result of all the stress from these emotions he had to manage. He was trying to stay occupied so he would not have to think about all that was happening around him and simply not taking a break from all his activities.

I went to my dentist to get my amalgam fillings removed. He had to numb my mouth, give me oxygen, and secure a rubber dam before he could get started. My mom went with me this time. Once again, I had covered myself with diodes and wore sunglasses. I asked the dentist if he could turn off the overhead lights and work with one of his lights only. He had no problem. I must say that I was hopeful and kind of surprised that I made it through the first procedure. I scheduled the next appointment for two weeks later. I was still doing the checkbook for the stores, paying all the bills, trying to be Mom, and figuring out how to heal my body.

Brandon was having his surgery at Carolina's Medical Center in Pineville, North Carolina. I went with him to do the paperwork and sign the forms. I was very bothered with the electricity and feeling sick. This was one of those times that I just had to put my mind over a matter. They rolled him back to surgery. My husband stayed there during Brandon's entire procedure and called me when he was in the recovery room, about to be released. I was upset that I had to leave him. This was one of those things a mom was supposed to be able to do. I sometimes felt cheated, but I had to stay focused. I knew his dad was there and it was okay. We brought Brandon home. My brother Ben and Nan came up from Charleston, South Carolina. Nan was going to be Brandon's private nurse. I was so thankful because I needed help, and I did not like the sight of blood. They stayed for three days and got us over the hump of post surgery recovery. I sincerely appreciated their help. Brandon continued to recover. A teacher came over to the house from school and tutored him with his assignments for the rest of the school year. He would finish his sophomore year at home recovering from knee surgery. The sad thing about this is that he would not be able to play football his junior year. He would be released from the doctor and might be able to play basketball at the end of the season for the following school year. I felt really bad for him, as I knew this would hurt his college scholarship opportunities.

I went back to the dentist and had the next set of mercury fillings removed. I noticed that when he pulled the last of my mercury fillings, my big toe on my right foot popped and my headache started to lessen immediately. I was lying there wondering if this was why I had so many headaches and why my big toe had been throbbing. I had lots of pain with my big toe for many years prior to this procedure. It had just gotten worse with time. Dr. Phillippi finished up and told me I had done great. He wished me luck and hoped I would find relief soon. I was actually feeling a little different. I was not sure what was happening right then but time would tell.

As my mother and I went outside, I told her what I was feeling and she was happy to hear some hope in my voice. What I had been noticing for the past year was that my teeth hurt and felt as if they were pulling and wanting to fall out. I suffered with bad headaches, but almost immediately my headaches had declined and that was a great relief. I went home and began to pay attention to the details of how my body responded to the mercury being gone from my mouth.

# JULY 2010
## The Verdict

My cousin, Rhonda, came for a visit from Nebraska. It had been a year since she had seen me. I was more swollen and felt sicker than when she visited the last time. I was trying to eat right. I did drink vodka to help numb my pain. I have never been a drinker. Actually, I never liked the taste of alcohol. However, it was the only thing I could find at this time to give me some form of relief so I could sleep a whole night through. I went back to Dr. Kanelos' office for my nerve pain, nausea, and headaches. Dr. Chris saw me and did the routine questions. Nan, my brother's girlfriend the RN, was with me. She asked many questions and tried to get some better answers using her medical knowledge terminology. Dr. Chris said he could not let me leave  there without providing me a prescription for pain medication. He saw how much pain I was in and how nervous I was feeling. He prescribed Gabapentin for nervous system pain, Valium for anxiety, and an anti-depressant. I did get them all filled. I was about to have a nervous breakdown and wanted to see if any of this would help my sickness.  I took the medicine for seven days with no relief, so I stopped. I would drink some vodka when I needed relief from my nervous sensation. I was tired of having to resort to alcohol for relief from the pain, but I was so miserable. Having been a bondswoman in my past and seeing people for who they are, I never judged people for their habits. My consuming alcohol to relieve my pain and suffering was unorthodox all the more because I was raised in a non-substance home. My parents were very strict and taught us natural remedies for any health-related issues. I quickly realized that people have to do what is best for them. I had to do what was best for me and if anyone could feel the pain I was in, more power to them if they wanted to take it on to resolve.

My cousin, Rhonda, went with me to court for my DWI. She had gone with me the previous year! She could not believe I was still going to court. I told her this was my 12th time. I decided that I was going to accept whatever the judge handed down to me. I was tired of trying to make them understand my sickness. The officers always showed up for every date and added their spin on the story. I was having an upchuck episode; looking insane, and had gained about 50 pounds. I had a lawyer who was not really for me. I was an outsider in the little town of Troy. I had no chance to reverse the charge. I tried and gave it my best shot to prove my EMF sickness was really the reason for the charge, because I was not intoxicated from one alcoholic toast. However, the judge found me guilty and sentenced me to unsupervised probation for one year; included a fine, suspended my driver's license to a privilege license, and ordered to complete 24 hours of community service. I had to attend an alcohol class for 16 hours in order to get my license back in a year. I was not too concerned with anything but the community service because I was so sick. I could not deal with electricity. I had to go downtown Charlotte a few days later to get registered and pay for community service. I was sitting there around several computers having an upchucking episode; begging the lady to hurry because I was sick and allergic to her computer's electricity. I was thinking how sad it was that I was dealing with a clueless society that didn't understand my problem. She started to hurry, as I got sicker and sicker. I got my community service assignment close to my parent's house in case I needed my mother for an emergency situation. I hurried out of that office to get some relief. Once I got home, I called the place where I was assigned to do the community service. It took two weeks for the person to return my call.

Meanwhile, I finally found an accepted medical facility and doctor in Texas who knew about EMF sicknesses. I called providing them my information. I told the representative how sick I was and did not believe that I could fly just then, nor even ride that distance in a car. I asked if they knew a place closer to Charlotte, North Carolina. I spoke with Sue and she said there was an environmental clinic in Charleston, South Carolina with a man by the name of Dr. Lieberman. So I contacted his clinic; asked questions about the place, the program, and how much it would cost. This was a four-week bio detox program. However, if I needed more time, those arrangements could be provided. I told his assistant that I wanted him to work with Dr. Rea in order to see what would work for me and my needs. They agreed with my request. Dr. Lieberman's specialty was chemical triggers, not EMF. Dr. Rea was the EMF expert and though there are similarities, they are still different. They had similar treatment programs within their own clinics. I decided to make an appointment since it would be six weeks before they had an opening. I scheduled my appointment for October.

I was so excited to finally find a place that understood environmental triggers. In addition, it was near my brother, Ben's house, which allowed me to save on hotel expenses. My finances had been very tight with the economy's crashing causing lower sales in our restaurants and me getting sick and unable to work. Due to the type of illness I had, I was going to need a lot of funds to afford my healing. My husband and I had to pay an employee to work in my place since I was not able to manage one of the restaurants in my condition. I was so excited about being able to go to this facility that I started calling my family and friends immediately. My cousin, Rhonda, was with me when I made the call and she was excited for me too. She had stayed with me for six weeks and she had to go back home. I was very grateful she had come to visit, and stayed as long as possible. I enjoy Rhonda's company because we are similar in many ways, but we do have our differences. She is the oldest cousin in our family on my dad's side. She looks much younger than her age and is very talented. Rhonda is a great cook and I certainly appreciated her help tending to me while I was down through this time. I do believe she was aggravated with me at times because she thought I was being stubborn when she asked if she could do anything for me. I would always tell her no. She had her own health issues with her back and I wanted her to relax in order to heal herself. I enjoyed her company since I didn't care to be alone all day and through most of the evening hours as well. I had now been in pain for 19 months. I had done a tremendous amount of research and was slowly figuring out EMF and ES sickness.

# SEPTEMBER 2010
## One Piece of the Puzzle

I needed to get the community service work done before I left for the environmental clinic in October. It was great timing that I had the amalgam fillings removed prior to this trip. The lady from the community service called stating I could come anytime and get started. I was so eager to get it over with that I went that same day. It turned out that I was able to do my community service in a perfect place, which was in the back of the Elon Recreational Center located in the woods. I had to pick up trash outside the facility and all through the trails and fields. Perfect for me with my problem, I thought. I went to the center in order to get a feel of the place and to see what I needed to do to fulfill my 24 hours of service. One of the jobs required was that I dust mop the gymnasium floor with one of those big push mops. Fortunately, the lights were off since it was really hot and they had no air conditioning. This was a real old place. I did explain to the attendant about my medical condition, but like others, she just listened and thought I was strange.

I was familiar with this mentality as I had been a bondswoman for 13 years. People judge all the time and assume its human nature. Especially if one is influenced by what they were taught. The first day I was able to get in four hours of service. On day two, I dust mopped the gym and proceeded to pick up trash outside. I called my mom and told her to pick me up at the end of the field. Understand it was September, 100 degrees outside, and I had a medical condition that was not yet recognized. I stayed at my mom's house for several hours then went back to the facility.

I was allowed to sign myself in and out without being detected or caught. By the third day, the male attendant allowed me to leave early. I was done with the community service. I really did about three hours of moderate physical work. I am not boasting. I prayed and asked the Lord to help me get through the task; place me with the right people so I could be done and get this behind me. As soon as I finished, I got a call from the Bio Detox Center (COEM) telling me that I could come earlier starting on Monday because they had a cancellation. My sub-conscious was telling me that if I got this DWI behind me, I could start moving forward. I was so excited and my spirits were lifted. I needed to go home and get the checkbooks finished; pack, and call my brother to let him know they moved the dates.

It was Monday morning and Mom and I started our journey to the COEM clinic in Charleston with Dr. Lieberman. I could drive, but I did get sick having upchuck episodes. I could tell you where all the Wi-Fi towers and high-tension power lines were without seeing them. My body responded to the exposure by upchucking. It was the same with a plane in the sky flying over-head. My body was super sensitive and on high alert with flight or fight mode. We got to the clinic and I had an interview with Heather, a medical coordinator with the clinic. She asked a series of questions. I had typed up a sheet with all my symptoms and problems. It was four pages long. She was taken aback by this list. She had to input all the information into the computer before I saw the doctor for my consultation. I told her I couldn't deal with the computer for even a short period of time. I started to upchuck like I had said. She quickly removed the computer and I waited for her to get done.

I did notice that I could feel some form of electricity behind my head while sitting in the room. Later on as I looked around into the other room, I could see it was the microwave that was bothering me earlier. I was shocked that an environmental clinic had a microwave. Furthermore, the entire facility had fluorescent lights. These two items are not good for anyone including the environment. Dr. Carol Benoit, who is a osteopathic doctor, came in first and went over my history all the way back to my birth. I was impressed because at this center they understood that a person does inherit some problems genetically; however, most health issues depend on what type of environmental triggers one grew up around. As an example, some of the questions Dr. Benoit asked me were, "Did I grow up in a new or old house? If old, did it have lead paint? Did I work around any toxic materials? What was my ancestry? What part of the country did I live in?" These types of questions help in evaluating an individual's health issues. I proceeded to go through the Q and A with Dr. Benoit sharing with her a photo album of my pictures that I put together dating back prior to my sickness. I wanted to prove how different I was before I got sick, and especially prior to my weight gain. Dr. Benoit said a lot of people claim weight gain as if I was not telling the truth. This made me want to punch her. I was very sensitive about my weight and I had just shown her pictures that clearly indicated something was going on that was not normal. I just took a deep breath and continued to answer her questions. She did notice I had one nostril that was not like the other one. This was from an accident I had when I was younger. I got hit by a hockey stick across my nose when I was three years old and my mom said I was very black and blue. My nose was probably broken, but they never took me to the hospital because my parents thought it was just a bad bruising. Also, my mother said I had fallen down the stairs, about 17 steps, when I was two. My baby tooth fell out with a really deep root attached. I said this tooth did not come back until I was nine years old and the color was a yellow. That indicated damage from the injury.

As we continued with more of our conversation, Dr. Lieberman came in to summarize my evaluation while proceeding to explain the treatments his clinic offered. I asked if he could help me correspond with Dr. Rea about my patient care and he said no problem, as they were great friends. I felt relief since I needed to be tested by Dr. Rea for EMF sickness. After my meeting, mother and I went by my brother's house in Charleston to let him know that I could not start the program the next day, as I would have to begin the following week on Monday. I had to get my finances together and see what I could do because the treatment program would cost thousands of dollars. I was feeling stressed about the financial burden since I did not have much money to work with and our insurance would not assist on the expenses. I decided I would write a letter asking for donations to be made to the center on my behalf so the charity would be a tax deduction for anyone who contributed. In the letter, I mentioned that this treatment program may help me in getting my life back. I thought this was a great idea, and I did receive a few donations through the center. Something was better than nothing, and I certainly appreciated the help.

The week went by and I finished what I needed to get done in order to be able to go in peace to heal. My mother and I left on Monday around five in the morning so we could be at the COEM center by eight when the treatment program started. The nurse, Elaine, showed me around the facility and told me what was going to take place. Nancy, another nurse, also came in assisting those of us who were there to begin our treatment program. I met the people who were sick: Joe, Tony, Kim, Zina, and Ella. They all had either chemical or mold triggers. I had electricity problems and the microwave had to be unplugged. The fluorescent lights had to be off, or I would have an upchuck episode. The nurse Elaine made a sign for the microwave which read, "Do not use while Beth is here."

Day One of the treatment program started with me taking vitamins and sorting the rest of them for breakfast, lunch, and dinner. I was prescribed 55 different supplements. Next, I started an oxygen treatment for 15 minutes while doing a simple walking in place exercise. That was all I could do because bouncing my head on the mini trampoline would cause my head to hurt. I could not do the stationary bike exercise because it made me more nauseated by the movement and noise. I then proceeded to sit in the sauna to sweat toxins out of my body. I was in for 20 minutes and out for 10. This went on until lunchtime as I was drinking plenty of my own ionized water. I brought my own water machine, hooked it up in the break room, and shared it with the entire center. My intention was to show Dr. Lieberman that this highly alkaline water was very important for healing. I also sold a few machines in order to help meet my financial obligations, as I have always been a working entrepreneur. We would break for lunch and we had to bring our own meals in to eat. I had a perfect, healthy salad and deer meat. Others had no clue about healthy eating or nutrition. Each of them would feel worse after eating lunch, especially when they had to get back into the sauna. Lunchtime lasted for one hour. I ate with Dr. Benoit every day. This was great as we had much in common with our knowledge about health. As time went on, I forgave her for not believing me about my weight. She had realized that it was from the EMF and not my diet because she observed what I ate daily. She said she had never met anyone like me and was not familiar with EMF sicknesses. I taught her a great deal as she witnessed the struggles I experienced with just the refrigerator turning on and off. She saw how I could not tolerate lights being on around me at all, or the airplanes flying over the building. After lunch, we would get back in the sauna. Twice a week, I would have a deep tissue massage and one lymphatic massage. On Wednesdays and Fridays, I was administered an IV bag containing Glutathione and Alphalipioc Acid. Every day I would have two hours of oxygen treatments using a porcelain mask in order to help the brain. We also had four hours daily of hot dry sauna therapy. I had numerous tests taken for the lungs, nerves, and any allergic reactions. All of this was a great experience and it gave me hope. I continued to have my EMF reactions and was looking forward to being able to see Dr. Rea for my specific issues as soon as I completed the COEM treatment program. When I would start having upchuck episodes, I took antigen drops that helped a little in calming down the ordeal. While receiving my oxygen treatments, I decided to re-read that book about EMF sensitivities. I realized Dr. Rea had been mentioned in the text. I totally missed that over a year ago when I first read the book. I now further understood more of what the different tests were designed to predict in one's health ailments and why I was having them done.

While I was at Dr. Lieberman's clinic, I received a return call from Attorney Nick from Columbia, South Carolina. I had called leaving him a message asking for assistance because Social Security Disability had denied me for the second time and Peper Law Firm recommended that I call him. Nick explained to me that he was not familiar with my type of condition and would need a written statement from a doctor who could verify my sickness. I told Nick that Dr. Lieberman could do this and that all the other doctors I had seen could confirm the illness. He said, "Listen, you need to understand what I am saying. In disability, with Social Security, you have to have a doctor that has the proper credentials and a proven test that verifies your condition. Just because a doctor says you have EMF sickness means nothing. I do not disagree with your condition, but if you want me to represent you, I have to have this documented proof. Okay?" I said I understood and thanked him for returning my call. After hanging up the phone with Nick I got very upset; started crying and worrying over the fact that I may not find a doctor who had the credentials or tests that were certifiable in meeting the SSI Disability requirements. Here I was suffering immensely for the past two years and it was not my fault that society or medical professionals were uneducated about the EMF sicknesses. So, I decided to call Dr. Rea's office to see if he had the proper credentials and testing to prove my sickness. I was relieved to find out he did. I now had to go see this doctor in Texas just for verifiable proof, and to learn more about EMF sickness complications. I called Nick, the attorney, back telling him I was going to see Dr. Rea on November 9, 2010, and that I would contact him once I got back home. I had finished my six weeks at COEM Center and was now ready to go to Texas.

# OCTOBER 2010
## Alive at 40!

Mom and I had driven to Charleston, SC every Monday morning to go to COEM. We stayed at my brother Ben's house during the evenings Monday through Thursday. My mom would drive to the center in the morning; drop me off, go to the beach, and later have dinner cooked for us all. She had balloons ready and baked me a birthday cake to celebrate my 40th birthday. We had a little party gathering during lunchtime with my new friends who were sick and in the treatment program with me, as well as Dr. Benoit, Dr. Lieberman, nurses Elaine, Nancy, and Jeff. My brother, Ben, and his girlfriend, Nan, surprised me by dropping by the party. I could not believe how I looked having my 40th birthday in a bio detox center! I was kind of sad having to deal with this and was thinking about life not being fair. I would just have to stay positive, keep my mind together and get better.

Sometimes, during the course of my treatment program, Ben would come pick me up after I was done. It was an 8 to 5 program because the facility did not house patients overnight. Everyone else had to stay in a hotel, but fortunately for me, my brother lived in the same town as COEM. After being in treatment all day, I felt like a wet noodle and my nerves were very fragile. I felt completely wiped out so while riding in the car back to Ben's I would feel very nervous and anxious. Some days, when I got to his house, I would have severe upchuck episodes that completely wore me out physically. I could never deal with the blue light shining on the satellite box for his TV, or the air conditioning unit running. I slept in one position all night on his couch with my head at one end. All the lights had to be off, and we lived in the dark when I was around. One particular day, my brother had a lamp on and I started to upchuck. I looked over and noticed it was one of those new CFL swirly light bulbs that have mercury in them. Those are highly toxic. I learned that CFL bulbs would cause me to have an upchuck episode and Ben immediately replaced the bulb with an incandescent bulb. That night I had to research the effects of CFL bulbs. I found a list of side effects including headaches being one issue. If a person comes in contact with a broke CFL bulb, that person could be seriously injured because the bulb contains mercury. This can release toxins in the air and if handled physically, could eat a whole through the skin. I got the gist that these were extremely harmful. I knew these bulbs were a problem for those of us who suffer with EMF sickness and I vowed to educate anyone I came in contact with about the side effects of these CFL light bulbs. The local power company had started giving these bulbs away for free to their customers in order to have them switch over from using the incandescent light bulbs. Some Green Energy efficient way to save the earth! I thought, what about us humans!

# NOVEMBER 2010
# The Lone Star State

The day came and it was time to board a plane to Dallas, Texas to see Dr. Rea. My brother, Ben, and mom went with me on this trip. Not knowing how I would react to the EMF in the airports and the ionizing radiation from being high up in the sky,

I was a little scared to take this journey. However, I was happy and excited about going to see Dr. Rea in hopes of healing my situation. I was leaving my children, husband, and all the responsibility of our checkbooks for the restaurants behind. I was the only one who knew all the details for the checkbooks. Family and friends often wondered why I was still doing these checkbooks in my condition without my husband's assistance. I must say that in most cases only one person in a business or marriage handles the checkbooks. I had always taken care of this duty. We obviously did not plan on me getting poisoned by power lines, so I chose to continue managing them. Besides, I did not have the patience or nerves to teach this to someone who was clueless and starting from scratch. It was easier for me to just do them and take my time, not having to go through the stress of teaching him. I was grateful he was physically running the stores and that there was no need for me to worry over the actual restaurants.

Well there with my dad dropping us off at the airport, I went into an uncontrollable crying spell and started upchucking as soon as we entered the area. My nerves were really being aggravated by the high levels of EMF. I managed to be positive and believe I could get through this no matter what happened. I was working on me and my project to help heal others in the world suffering from similar circumstances. We got to the body scanners and I presented my doctor's note stating I could not go through the X-ray machine and must be patted down. This seemed to irritate the officer, and I thought to myself if you only knew what I was going through. They finally got someone over to pat me down. I was having a major upchucking episode. They did ask if I needed a trashcan. One would think that upon witnessing me go through an episode like this they would have moved quicker. However, I finally got through that checkpoint and took some antigen drops to help calm me down. We walked to our gate where we were to board the plane and sat down. I sat still and as calm as I could, not wanting to make a scene drawing more unnecessary attention to myself. These upchucking episodes really take a toll on my body and appearance.  They are sudden and compulsive, causing me pain all over my body.  It's a chore to keep myself together while going through one of these, but I try to remain positive about the situation.

As we boarded the plane, I started upchucking again and I was trying to be quiet about it thinking that if these people only knew what I was going through. This was horrible. How could any scientist deny this? And what about the power companies? The plane got ready to take off. We were airborne and I continued to have a massive upchucking episode. When the plane got high enough up in the air, I settled down. I sat real quiet and remained positive. I couldn't help but wonder why EMF/ES was so secretive and silenced. It's very real and happening because I was suffering from being exposed unknowingly. I knew that if I was suffering, then there must be many others all over the world suffering as well. I had to find a remedy. Something just had to be done about this sickness. I understand that society could not handle an epidemic of fear so finding remedies had to be found. I knew my body could heal. I just had to find the right ingredients. It was time to land in Dallas, and as we descended I started to upchuck. I was thinking the whole way, I can do this, and we are almost there.  We landed and hurried to get our luggage; then caught a taxi to take us to where we were staying. We checked into our room that was supposed to be EMF safe. We unpacked and went to the Environmental Health Center, Dallas to see Dr. Rea. I was excited to meet him, but I had to complete my paperwork first.

The moment I had been waiting for, to meet Dr. Rea, was surreal. As we waited to see the doctor, his assistant, Trep, came in and we all introduced ourselves. He said he had heard a lot about me from Dr. Lieberman and that he was going to help me. I told him about the power lines and how I had been suffering through so many symptoms.  I had it all written down in a four-page report. I could not really remember the last 20 months from memory so having a journal proved to be very helpful even for the practitioners. We went over a plan of action and got started immediately. He also told me that the upchucking I was referring to was known as retching, so I began to use that terminology. We started with a test for my sympathetic and parasympathetic nerves. I had numerous allergy tests done. Heavy metals were a problem, including mercury, tin, copper, steel, nickel, and titanium. He created an antigen shot that would be mailed to my home in order to prevent complications with the airport staff. I had a nuclear brain scan, EMF test, as well as the thermal breast and body scans. I had several tubes of blood work done, which was very difficult because my body acted as if it did not want to release the blood through my veins. This was very painful and upsetting considering the day I already had with my health.

The technician had to repeatedly stick me in an attempt to draw blood. They ended up having me to go sit in the sauna in order to help my blood vessels expand. Unfortunately, the sauna did not help and my arm felt like a pincushion. Dr. Rea explained to me what was happening. He said that my body was in a very sick state and that it refused to receive or release due to the damage. This process is referred to as the flight or fight mode and my body reacted in high security watch.

The days went on and the results came back from all my tests. I told Dr. Rea that the apartment we were in was not EMF safe. I did not sleep well at all. There were several problems with the room. One was the digital alarm clock, which we unplugged, and two, there were huge power lines outside my window. Therefore, I was under constant electrical attack. I wanted Dr. Rea to know my body was so sensitive that it acts as a true tester for EMF effects. All the research I had done was verified by what I was physically encountering. Furthermore, my arms were severely bruised from the punctured wounds endured by the nurses inability to draw my blood effectively. Both my arms were punctured with deep, black, blue, and purple bruises. The nurse had really hurt my arms. In fact, my mother, who was a former nurse, said she had never seen bruising or coloring from blood work look this bad.

According to the test results performed, Dr. Rea told me that I had electric magnetic hypersensitivity rendering me allergic to 60 hertz of electricity. That is the hertz frequency standard for electricity in the USA. Furthermore, when exposed to this level of frequency, my blood pressure rises while my eyes become dilated. In addition, I did suffer brain damage in the locations where logical thinking and problem solving take place. Dr. Rea wanted me to go see Dr. Diedrikson in town before I left Texas. This doctor had a great test that would be useful in my behalf for lawsuit purposes. This test would verify the brain damage in an image in order to prove I was not crazy in my pursuit about my case. Dr. Diedrikson, a neurological psychologist, is a great specialist and highly respected. She was fully able to present her facts if I decided to proceed with a legal case against the power company. The extensive blood work covered an STD and HIV Aids test, which came back negative. I was put in the position to cover every angle possible with regard to studies and a lawsuit. My thermal scans indicated that my endocrine system had problems along with the nerve tests, proving my nervous system had been damaged. Hearing all these reports only confirmed what I already suspected about my condition. I felt sad hearing the reality from a medical standpoint, but relieved at the same time because this was the proof that I needed and had been searching for since April 2009. These battery of tests solidified that I was neither crazy, nor hypochondriac. There were very serious issues going on with my health that needed to be addressed in order to save my life.

I asked for the science behind my brain damage and physical condition. As odd as this may be, it's the honest explanation about the ordeal. When I was working in the yard building our landscape design, for many hours I was in close proximity to the power lines and the amalgam fillings in my teeth were acting as metal conductors to my brain. My body went into high stress producing histamines and Cortisol at extremely high levels. Eventually, this constant bombarding was too much for my body to process and eliminate causing my blood vessels to leak producing enormous swelling throughout my system including the horrible belly fat. Upon hearing all this, I felt as if I were having a brain overload moment because that was a lot of information to absorb especially considering my emotional sensitivities to weight gain and my appearance as a whole. I have always been conscious of my appearance trying to look my best, being presentable as I worked in the eyes of the public. Also, I cared about my quality of health. Who wouldn't want good health? This all was very emotionally hard to endure.

Dr. Rea said that I was obviously very smart or I would never have been able to perform at the level I had. Most people would have given up and quit. He stated how impressed he was with all the knowledge I had and never before seen someone as bad off as I was in this condition. I shared with him that I had read in a law book that in order to win a case the person should know their story better than anyone else. So, I took that literally to heart. He told me the electricity in my body had been readjusted. He believed that one day I would get better, but had no idea how long it would take. He compared my situation to getting an adjustment at the chiropractor's office. When the chiropractor adjusts some-one, the body will either hold it, and the person gets better, or not and they will have to continue going back until it takes hold. I did feel some comfort. Although I was still in pain, I resolved to take one day at a time. After a long day, mom and I got back to the hotel room; ate dinner, and then went to bed. I had an appoint-ment with Dr. Diedrikson the following morning at eight sharp.

Mom and I got up early; ate our breakfast, and I took all those vitamins as we waited for the taxi driver. It was hard on me because the power lines on the street outside my suite bothered me. The taxi came and our drive was about a 40-minute ride. I was sick the entire time, but was dealing with it optimistically. We met

Dr. Diedrikson. She was very stern in her approach. She stated she had seen everything and that there was nothing she had not seen in all the years of practice. Therefore, don't think she can be tricked. Well, don't try me I thought. You have never even seen me before and have no clue how I have suffered. Her attitude and approach took me back a little. I assume patients came to see her pretending to be sick or mentally ill. However, I would not wish my problems on anyone. We spent the first two hours answering questions that she addressed to my mom and me. I had made the habit of telling every doctor about my DWI charge in case I ever went to court with a lawsuit. She asked some details about the incident and I answered. She said that with my condition my body could not digest properly and it really would not matter the amount I consumed. My looming question about that DWI was finally answered. I immediately thought what a joke DWI was and it must have been a reason for me to go through the experience. The DWI and the court system made my life story even more interesting. I just had to chuckle on the inside rehashing that entire process.

At the conclusion of the interview, Dr. Diedrikson told my mom to come back in seven hours to pick me up, but to call first. We quickly started to do the testing. She soon realized that I had major issues. If I flipped paper, I started to retch. If I tried to put a puzzle together, I would retch and start crying. I could not do simple math like adding numbers or complete any type of mathematical equations. When she read me a story, I had very little comprehension in being able to repeat back to her what was read to me. I started to piece things together and remembered when I would work on our checkbooks or any paperwork, I would usually retch. I had to use a calculator and write everything out constantly because I could not remember what I had done. This explained why I got confused about where I was and if someone told me a number. I could not process the information in my memory causing me to forget more than two numbers at a time. If it were three, I would forget all of them. My husband would get frustrated with me each time because he would tell me something and I could not remember what he had told me. This information further explained why I was feeling as if I was in delay mode. When someone would tell me something, it took time to register. I was getting a lot of my unanswered questions finally answered by taking her tests. I was in shock about the condition I had succumbed to over the past two years. I couldn't even perform the simplest task. Two years earlier, I was able to add several numbers, columns, give my employees assignments, plus answer questions all at the same time. I was really good at multi-tasking. Now here I was and I couldn't even add some simple numbers or repeat a three-year-old story back. I continued to take more tests with this doctor and the time just flew by so quickly. It was now time for me to leave. While I was taking the test, Dr. Diedrikson said she had never in all her years witnessed anything like this. She told me that with my determination to get through pain and to do what I was asked to do, I would figure this out and be able to help many people.

She said I would need to write a book after I was healed. She thought I was a unique individual and told me that she appreciated her time with me. The taxi driver came to take me back to the hotel. My mom and brother were waiting for me to arrive in order to pay the taxi fare and help me get to our room. I came in and just started to weep profusely. It had been a very difficult day. The pain and the reality check that I really had a lot to overcome and it was not going to happen any time soon, caused me to break down. I was stunned how the power company built the power lines so close to my house. They knew what the bio effects were on a human life, but continued to deny the facts rendering many people to get even sicker. I was just in the forefront of this sickness. I should have been warned before we purchased the house that I could not live there especially with amalgam fillings. Oh well, enough of that negative talk. I wanted to pull it together, keep on moving forward and now get ready for dinner. I also needed to appreciate that I had received this much knowledge and verification from board certified medical doctors.

We were able to connect with an old family friend, named Lance, who had grown up with us and now lived in Dallas, Texas. He was not far from our hotel. So, he came and picked us up to go out for dinner and get caught up on the past twenty years. We had a great dinner and enjoyed seeing him. Besides my issues of being in a restaurant, all went fairly well. I always had to figure out where I could sit. We tried one restaurant and had to leave due to my body not being able to handle the high EMF levels. I was sure that Lance was wondering what in the world was going on with me. He had his own chiropractic business in Texas, Vista Ridge Chiropractic. We enjoyed our time with him so much that we decided to go to dinner the following night and meet his three girls. I had one more day at the center to meet with Dr. Rea before leaving to return home.

On the last day we summed up the things that I needed to do in order to heal. Dr. Rea informed me that I did live in a highly toxic area due to the Catawba Nuclear Station and the former Bowater paper plant. In addition, having a chlorine pool near power lines is a lethal combination to the human body. I should do energy medicine such as reflexology as well as trying to limit my exposure to stress by all means. If I could ground by being barefoot outdoors allowing my feet to come in contact with the earth's natural grounding forces, this would be very beneficial similar to the time when I went to the beach with my brother Ben. My body was grounding at the beach and pulling the electricity out of my body as my feet turned burgundy. They did have an indoor grounder that Trep, his assistant, made and he would mail me one to use at home. He told me that he would be happy to talk with my lawyers and to help in any way he could, if needed. I asked him to sign my copy of his book that he wrote about building an EMF and ES safe house. He said he appreciated my knowledge and positive attitude. He couldn't give me a timeline but he assured me that I could heal from this situation. He also created me an antigen shot that I would have to administer to myself three times per day that the center would mail to my home. The shots would help me not to retch.

Mom and I left Dr. Rea's office and went back to our hotel to wait for Lance and his girls to go out for dinner. Meeting Lance's daughters was great as they all were really nice, cute, full of charm, very smart, and athletic. I went to school with Lance's sister, and my oldest brother, Ben went to school with Lance. It was nice to go back in time and just reminisce over so many fun stories.

It was Saturday and time to leave. Another childhood friend, Donna, and her family lived nearby, so we were waiting on them. We also had not seen her in several years. We were going to lunch, and then they would drive us to the airport. We went to lunch and I did my usual routine of trying to get the best spot in the restaurant for my body. We ordered; ate and visited. Donna was younger than me. I had grown up with her sister and we had a lot of great times together many years ago. It was nice seeing her along with her three children, and the way she interacted with them. She enjoyed many blessings, including a great husband. I was so proud of her. It was time to go to the airport and get back to my family. Donna drove and the closer we got to the airport, the more I started to retch. I had to be strong, and again get patted down at the security checkpoint as well as get to the gate. I was retching through the whole process and crying from my nervous system being on high alert from being damaged. It felt as if I was being electrocuted on a lower level. My nerves felt like thousands of needles were sticking me all at once. I was nauseated and my head hurt. However, the intensity was not as high as it was when I had the amalgams fillings in my teeth.

Our plane took off and I continued to retch until we got high up in the sky. I sat very still absorbing what was told to me over the past week. I was grateful for family support and discovering some medical help that was knowledgeable about EMF/ES sicknesses. Dr. Rea was the only doctor in the US that I knew of who could diagnose electrical magnetic sickness, even though Dr. Pan in the center actually performs the series of tests. It was time to land and I started to retch again. I was home free and I knew I could get beyond this episode with putting the flight behind me. I was excited to return to the COEM Center in Charleston and update everyone on my prognosis. I was looking forward to discussing my experiences with Dr. Lieberman. So, I got home, and caught up on the Subway restaurant checkbooks, plus all the mail that I missed. I now understood the retching I experienced from handling papers and getting upset, which had to do with my nervous system.

My family was excited to have me home and hear about all my experiences as well as any information that would make me better. I was able to plug my grounder in the outlet which was crazy looking. It was a short copper wire connected to a copper bracelet which was wrapped hooking to a plug with the grounder prong part only that had to be plugged in to an actual electric outlet socket. I was to put this grounder around my ankle making contact with my skin while being plugged to an outlet. This grounding really helped me by acting as a faraday cage around my body allowing me to feel free and protected. However, with only 10 foot of wire I was totally attached to an outlet, which was encouraging, but limited my mobility.

Monday came. Mom and I had completed all our duties since returning home over the weekend. We were returning to Charleston to see Dr. Lieberman. We arrived at the clinic and everyone was happy to see me. Dr. Lieberman said he had missed me because I was a big help to them at the clinic. My personality is to always be helpful to others and see the positive side. I am not a negative person and I try to bring the best out of everything I do. My new friend, Tony, and I spent many hours in the sauna having conversations about many different things. On this round of therapy, Dr. Benoit would be starting the Bio Detox for my remaining four weeks. I was able to talk with her for many hours and gain further understanding of the in-depth science about my body. This was a real treat for my project in helping me piece things together. I still searched for answers. I wanted to know why my breast leaked, especially since I was not pregnant or breast-feeding. Dr. Benoit told me that it meant my pituitary gland was in big trouble. I wondered why the OBGYN doctor did not know this since he dealt with the endocrine system all the time. I shook my head in disbelief.

A few weeks went by and the Christmas holidays were approaching once again. I had to call Nick, the disability lawyer, asking him to order my paperwork. I also needed to make an appointment to appeal my disability case for the third time. Nick scheduled an appointment for the middle of December. I called the Peper Law Firm so we could discuss my legal case together. I went on the Friday before Thanksgiving. Mark said they had con-tacted many law firms and was unsuccessful in securing one to help with my case. He said I had a case but needed money upfront to even begin the paperwork process. Mark and Dyllan were both fairly new in their law practices so funds were limited for them as well. I never had total peace that it would go all the way, which was disappointing, but not surprising. Mark said they would give me any research they had gathered along with all my records of information back. I was curious to see what they found during their research. He said he was waiting to hear from two other law firms and if they said no they would have to withdraw from my case against the power company. In the meantime, I continued to go back and forth to COEM Center until the week before Christmas. That Friday, I had my appointment with the disability lawyer, Nick Callase in Columbia, South Carolina.

# DECEMBER 2010
## Disability Deadline

Nick looked over my paperwork and realized the appeal had to be filed by 5 o'clock that day. I started stressing and he said, "No problem, no worries". I asked him if he could unplug his fish tank and turn off his lights as they were causing me to get sick. Thankfully, he didn't mind. He began to have problems with his computer and stated this had never happened before. I told him it was most likely because of me. This is another one of many EMF problems; my energy interferes with computers. We continued to talk while he gathered all the information needed and faxed the paperwork himself. I gave him all the research I had so he could become familiar with EMF sickness. Nick ordered all my medical records and told me this would be a challenging case since there were no previous cases like it we could use as a guide. He said he would do his best for me, and I should forward any information I received that would help. He would also talk with Dr. Rea; get his perspective, and we would go from there. Nick did state this process would take another year or two. I said, "What? Are you serious?" I was upset that this was how the system worked. Only if you were blind or had a major known illness would they speed up the process. If I did not qualify I would have to stay positive and hopeful. I told him this case was not just for me but for the EMF community and all who suffer. I wanted to make a difference and establish that this disability is real.

Another Christmas arrived, but this time we had no tree and no decorations. Everyone was simply worn out from the past year. My situation had taken a toll on the family. I did try to stay positive and not tell how I really felt in order to just keep the peace. Our finances were extremely tight. This causes major stress for anyone facing an expensive illness.  Especially, even more so, when you really have your hands tied not being able to work and produce an income. I kept quoting my scripture verse and praying. I limited my conversations with negative people or ones who tended to stress me. I realized I could not allow my husband to bother me by taking his comments personally. I understood he did not know what to do and was scared. These types of emotions can make a person respond in meanness. I knew he loved me and used the only coping skills he had been taught. If a person avoids and denies a problem then the situation is not really happening in their mind. It was not pleasant for me to endure but I understood. I was realizing that if I allowed him or others to upset me, it affected my nervous system and I would retch. My brothers' friend, Jeff, made me a new grounder with an orange extension cord, a big gadget for my leg, and 25 more feet, which allowed me to maneuver a little further in the house. I certainly appreciated his time and effort. The one that Dr. Rea's assistant made for me was not holding up well. I continued to stay on my bed and operate my life from there every day. I could not cook at all. Ovens still made me sick, as did dishwashers, washing machines and dryers. I was giving myself antigen shots, taking my supplements, and drinking a lot of ionized water. I had healed a little. I did not want to drink vodka, nor was I on any medication. I waited and continued to do more research trying to find someone who could test the house for EMF levels and micro radiation waves. Obviously, I had to have someone who was certified for legal purposes, if necessary. My brother, Barak, came to visit me, along with my niece, Emily, and nephew, Kavan. I had not seen them in a long time. Barak had been talking to me by phone, but really did not understand what was going on and was surprised by how I looked. It was nice seeing him as we caught up on all that was going on in each other's lives. I did have an appointment scheduled with Dr. Lieberman in January after the holidays, which gave me something to look forward to in the New Year.

# 2011

# JANUARY 2011
## AH HAH Moment!

January has arrived. The holidays were over and I had a lot to do from my bed, which became my office. I had a telephone follow up appointment with Dr. Lieberman. I had my questions written down in order to be efficient and have a productive call. However, he made a comment that did not pertain to me. I knew he did not know he was talking to me for the first five minutes of that conversation.  Nothing was accomplished in that phone call except me receiving a bill. When I hung up the phone, I realized this doctor was not focused on me during that appointment. He was super busy and had numerous patients to attend. He gets up every day and does what he wants to do. I appreciated all his care while I was at COEM, but it was time for me to move forward on my own. I had no choice but to really figure out this problem for myself, or I would be lying on this bed until I died. This was a new day for me and a huge light bulb going off—AH HAH moment!

My daughter was still home from college. I needed her help in sorting our Subway restaurant receipts in order to get our taxes done. I got very sick flipping papers and it would help having her assistance. She also helped in adding up the totals. No one can even imagine the pain and suffering I went through while still having to perform. I was calling the power company and telling them I was sick. I had been keeping this quiet until I was sure of what I was talking about and had medical documentation to tout. Furthermore, I had a guy coming to test my house on March 23, 2011. He has the equipment, but is not certified for court or legal purposes. I had already caught on to the silent propaganda and living in this much pain was my truth, so I figured I have nothing more to lose. I decided to get my thoughts together and call back in a few days.

Since I managed the checkbooks, I was able to discern this holiday season was the worse one we had been through, and our money was tighter than ever. I told my husband we could not meet all the bills and that we needed some money. I was also thinking it was time for him to bear some of this burden with me. He literally sat down on the bed and said, "Just tell me when we are going down". He went on to comment about the Hummer car payment along with the cost of gas to fill the tank, and that if I did not have those expenses, things would not be as bad. However, he missed the principle and quickly jumped at turning our problems around to be solely my fault. Even though he made a valid statement about the expenses, he failed to recognize the fact that the vehicle was bought before I ever got sick. This present situation was out of my control. I needed his support and assistance, not fault finding or making excuses. During our conversation, I noticed how I started to feel in my body. Then I went into a retching episode because my nervous system reacted to his negativity.

I could not tell him that I was already planning on trading my Hummer for a less expensive vehicle because if my plan to trade the Hummer and our daughter's car did not go as planned, I would hear more of his negative comments which I really was too sick to endure. I allowed myself to react to his negativity causing my nervous system to hurt so bad I felt like I was going to die. I lay on the bed, retching repeatedly trying to calm myself down while crying over the pain. Our daughter, Brandi, came in the bed-room and knew something was wrong. She said, "Mom, just let it go and let's go do what we need to do." Unfortunately, my hus-band was not very supportive or helpful, making matters worse with his negative approach and mindset. When people are physi-cally ill they need to be around positive people who don't produce more stress on the body.

Once again, I pulled myself together and decided to tackle the mission of replacing our vehicles, which needed to be done that day. My daughter and I called to get the value and payoff on the Hummer since we needed to be able to replace my car. Hers was already paid off. As the day went on I decided that I did not need a car since I was too sick to drive. Being that I was unable to sit in a dealership; test drive cars, and my daughter had to return to college, we called around stating exactly what we needed over the phone and waited for a return call. Thankfully, the guys at Hyundai South Boulevard Charlotte worked us a great deal so I could pay off the Hummer balance as well as have a brand new car for my daughter. This was a wonderful blessing that I was able to get two new cars for less than the total payment of the Hummer. Brandi and I left Hyundai with a brand new car, minus one big car payment. In addition, we would be using less gas and able to save on necessities in this area as well. I felt relieved and reported to my husband that I had fixed a multi-faceted problem by being able to relieve my debt with the Hummer. Brandi would have a brand new reliable vehicle providing great savings on gas.

I went to bed that night and had a vision that we needed to sell one of our Subway restaurants. The sales were not good at the one location and my husband, Johnny, could not be in two places at one time. Furthermore, I needed money for my treatments and supplements. The next morning I shared with Johnny this idea to further help relieve the financial burdens we were facing. Although he didn't really want to; he agreed. I called two other Subway franchise owners we knew asking if they were interested in buying one of our locations. The first person said no and the second one said yes, as I believed he would be the one that was interested. He was kind and said he was honored that I had called. I told him what had been going on with my health and how we needed to ease the situation by selling one of the restaurants. He also agreed to keep the purchase quiet as I did not need any employees quitting before our agreement and pricing was finalized. We submitted our paperwork to Subway and waited for further instructions. We were able to settle on a price over the weekend, which moved the process along quicker.

I called the local power company asking for the person who handles cases of people who had become sick from power lines. I was transferred many times, leaving numerous messages with different people. I videotaped myself retching while having the conversations, as this was very emotional and stressful to experience. At 5 PM that day, I got a knock on the door. My son answered and said, "Mom, someone is at the door for you." I wondered who it could be as I looked really rough. I went to the door and found that it was an engineer with the Health and Safety Department from the power company. He introduced himself and asked if he could help me. I said "Yes, I am sick." He replied, "I am not a doctor." I said, "I am not one either, but I'm sick and I know that your power lines have made me this way." He told me there was no scientific proof that power lines make humans sick. I said, "Well, I am sick and my body is showing proof." I told him that my whole life had changed. I couldn't do anything and had to constantly use this grounding device as I pointed to my leg. He was very argumentative and acted in disbelief. He asked if he could test the house. I told him no, as I had already had it tested and I didn't trust him. It didn't matter anyway because I had proof that I was physically sick from the exposure. That's one reason the power company bought our home because one of their representatives had tested the power lines during the summer of 2008, which was after additional lines had been added in 2006. The added power lines created even more electrical pollution on my property producing higher levels of radiation.

I had done my research and I knew a lot more than he thought. In my frustration, I went on to tell him that if the power company did not help me, I would tell the world. I further told him that if they tried to kill me or have me killed, many people would suspect his company. My situation and case was well-documented with many people all around the world. He did apologize for my problem and said someone would get in touch with me shortly. I thanked him and shut the front door. When I walked past the kitchen, I saw him in the backyard (in the snow) beyond the pool, trying to get a reading with some equipment. I quickly snapped a picture through the window for my own personal evidence.

I then went back to my bed and tried to calm my nerves. I received a text on my cell phone from my brother-in-law because I had mistakenly dialed their number earlier. I had not seen or heard from my sister in three years due to a family dispute. Her husband texted me and I responded with, "I would like to see my sister in case I die soon. I have been poisoned by power lines and have been very sick." Not expecting that comment, he immediately texted back and asked if I wanted to see her. I answered, "Yes". They were coming to see me the next day, so I went to bed earlier than usual because my sister was coming early in the morning as she has always been an early morning person.

I got up early and dressed, as I was expecting my sister to arrive at any moment. There was a knock on the door. When I opened it, both my sister and her husband looked surprised. Then my sister asked, "What is going on?" We hugged and I cried. What I was experiencing was a nervous sensation all over my body that caused me pain whenever I felt happy or sad as a reaction to the chemicals that are released while feeling those emotions. My body goes into a fight or flight mode because I had been in chronic stress for many years. My nervous system was damaged and that is why I tried to stay calm and isolated in order to prevent more stress. We talked and caught up a little. Our oldest brother, Ben, dropped by while en route to our parent's house. He had not seen them in a long time either. They hugged briefly as he was in a hurry only staying a few minutes to say hello. Well, the phone rang and it was a representative with the risk management department from the power company. He asked numerous questions and said he would get back with me. This was awkward to receive a call on a Saturday and the timing was quick. My sister and I continued to talk and catch up on life's happenings. She did offer for me to come to her house if I wanted to get a break from home. I told her that I would let her know. They went on their way and I went back to bed, trying to recoup from the triggers of excitement I had been through that week. Then another thought come to my mind: I should get rid of Brandon's Hummer 3 and trade it for a Hyundai with better gas mileage and a new warranty.

I went on Monday and took care of this. I did call first so they could get the paperwork going since they already had our information. I put this vehicle in my and Brandi's name in the event something did happen to me, God forbid, it would not matter. Brandi would have to go back to the car dealership and sign, or they could mail the paperwork to her making one more task accomplished. I was also able to include oil changes in the payment. This did take some of the burden off me. We skipped two car payments that month with both vehicles being traded. I had to figure things out quickly as it was the stress that I didn't need. Odd as it may sound; I just wished I could actually be sick in peace. It must not have been meant to be and it did make for a better testimony. I tell these details to encourage others. Sometimes, you really have to regroup and get back down to basics in order to survive. Now time had gone by, and I received a letter from the representative with the power company. The letter stated that, after their investigation, they would have to deny the claim. Go figure. I knew they were not going to admit to any wrongdoing. After working with the attorneys and reading all the articles I could find related to my issues, I decided to write a rebuttal letter with the help of my dear friend, Omega. I understood that some scientific researchers state that the EMF/ES sickness claims are unknown as they have easy, weak, naked, and junk research backing the claims. Well, my great friend helped me write the letter below addressed to the CEO of the power company including a phrase that he stated in an interview. (I have redacted the names of the other parties.) This letter was one of my last attempts asking for help from the power company considering their power lines almost took my life. I decided at this point to continue trying to find a solution, seeking additional medical help while remaining positive, believing I could beat this situation.

219 Oxford Place
Fort Mill, South Carolina 29715

REGULAR AND CERTIFIED MAIL
70070220000296818887

January 24, 2011

████████████████

Chairman, President and Chief Executive Officer
██████████████

Post Office Box ██████
████████████████████

RE: Personal Illness from EMF at residence
Report Date: 1/6/11
File #: 546992

Dear █████████████

I am in receipt of a letter dated January 19, 2011 from ████████████ Lead Risk Analyst, ████████████ ████████ in which he informed me of the denial of my claim referenced above. I must say I was disappointed to learn of his decision. Disappointed, particularly because I strongly believe in going directly to the source with attempts to settle

issues/concerns before they become bigger issues.
I contacted ▇▇▇▇▇▇▇▇ seeking assistance in dealing with the affects of my personal illness that I
firmly believe occurred from exposure to the power
lines surrounding my backyard.

Why am I contacting you? Because, I intend on making every effort possible to obtain help for my illness. ▇▇▇▇▇▇▇▇ in a statement attributed to you:
"I'm an optimist. I think there are solutions to
problems — maybe not perfect solutions today, but
over time, solutions will improve. I think the
probability that we'll get good solutions to cli-
mate change — solutions that benefit both the
planet and industry — is higher if we face the problem now than if we bury our heads in denial. If
you're constantly trying to define the problem, or
deny it, or dispute it, it gets increasingly difficult and costly to develop a good solution." I
trust this statement is true and this is what you
honestly believe. If so, I implore you to read my
file and reopen it for further investigation and
eventual resolution.

▇▇▇▇▇▇▇▇ stated in his letter, "…we have been
unable to find any indication that ▇▇▇▇▇ equipment
has caused or could have caused any of the health
problems you've informed us your doctor has diagnosed." I was not surprised by this statement and
even understand how it differs from findings of EMF
experts. My dilemma is I have EMF poisoning! I did
not have EMF poisoning; neither did I exemplify any
of its symptoms, prior to two years ago. EMF is most
prevalent around power lines. It makes sense my
illness began from exposure at my residence. The
same residence ▇▇▇▇▇▇▇▇ purchased from us.

Although I appreciate the empathy expressed by ████
████████ employees, empathy does not pay my medical
bills, nor does it protect my family members or me
from any further problems associated with exposure
to EMF poisoning.

Close friends who have watched the direct devastating effects of this disease on my life and me for
the past two years, continue to try to convince me
to begin a public campaign for support. I chose to
contact ██████████████ directly. I trust I made the
right decision. Please consider my request to
reopen this claim. I look forward to hearing from
you soon.

Sincerely,

Beth Sturdivant

Cc: ██████████████ Lead Risk Analyst
Certified Mail:   7007 0220000396818870

Chris's mother passed away during this month. Chris grew up
with my family and we were all good friends. I went with Ben and
Nan to visit his family. When we arrived, most of the people there
had not seen me since my dad's birthday party seven years earlier.
Everyone saw my latest grounding device I had just gotten from
my brothers' friend, Jeff. It was large and crazy-looking, but it was
like my oxygen tank in order for me to be able to endure electricity and guard my nervous system. Our good friend, Chris, said to
give him a few days and he would come up with something different. Chris is very skillful in building things.

While visiting with his family, I realized most people did not have a clue about how much added EMF is required for all our modern day conveniences. Among these items are dimmer switches, clap on lights, drop cords that are used as surge protectors, Wi-Fi, CFL light bulbs, ceiling fans, digital electronic control boards on appliances, and the list continues. The smaller the house, the higher levels of EMF gets trapped inside which causes worse exposure. I could not deal with his house at all. We had to leave earlier than expected. Chris called me a few days later and told me he had come up with this gadget that was a lot lighter and the cord could be as long as I wanted. My mom picked it up to see how it tested on my body. It worked very well, so I started to use this latest grounder. I wanted to show this latest grounder to Dr. Lieberman in hopes he would sell them to patients needing one. I decided to have this prototype patented because it was well designed and helped protect my body better. Furthermore, this would provide Chris a great token of my appreciation.

I was able to send the required paperwork for the sale of our restaurant to Subway headquarters this month. I continued to research more information about EMF, searching for anything related to healing while trying to stay stress-free. My sister had been calling checking on me and we would get together once a week. I was not really driving that much now so she would come and get me. I spent a couple of nights at her house visiting with my niece and three nephews. I enjoyed seeing them for the first time in three years. I had to stay plugged in with the grounder so I would not retch. I noticed they had a lot of power lines near their house. When the wind blew, a lot of the dirty electricity would be circulating in the air further irritating my condition. This would make me retch even while I wore my grounding device. However, I continued to visit my sister once a week for a while. I kept managing our Subway checkbooks, preparing more documents for the sale of the restaurant, and taking 55 supplements a day, along with my antigen shots. It was now almost February and my friend Caroline came across information for a facility that used the same water ionizer machine that we had, and they worked with individuals who had brain injuries or similar issues. I looked at the website, decided to call and left a message.

# FEBRUARY 2011
## Return Call

I was impressed that I got a return call within the hour from a Dr. Allen who had a doctorate degree in nutrition. I explained that I had suffered brain damage and had electric magnetic sickness. I went on to say that I would get sick just from flipping papers, and I had been to many doctors, the best of the best, dealing with environmental illnesses. I asked her directly what made her think she could help me. She explained to me that my blood brain barrier had been compromised and my body was lacking in Glutathione production for starters. I asked her to spell this word so I could write it down and look it up later. I still struggled with trying to remember things long enough in order to write down any kind of directions while working to comprehend the new information. I had major delay with my comprehension and memory. This was all due to my brain being damaged. I inquired how to get this Glutathione supplement and she told me that I could order it through her, which I did. I asked her about the treatments the facility offered for someone in my condition as well as the cost. In addition, I wanted to know when and how long the sessions would need to be scheduled. She decided to send me a packet of information with the forms that I would need to fill out and get back to her in order to proceed. After hanging up the phone and reviewing the new information she shared with me, I got very excited because I felt I was onto something that would take me to a higher level of healing. I called my mom right away and told her about our conversation, along with Brandi, Brandon, my husband, and Caroline. This new information really gave me some much-needed encouragement. I thanked Caroline for sharing that website link with me. I had a lot to do to prepare for this trip if I was going to attend a session of treatments. Just thinking about the amount of money required was very stressful. I felt guilty about spending our funds on treatments at times. I had to keep telling myself that it was an expense we had to incur since I had no control over the costs involved. Nor did I know how to prevent this situation of me getting EMF sickness, which could only have been prevented by us not moving into this house.

My son was now able to return to playing basketball after being released by the orthopedic doctor who performed his knee surgery a year ago. This meant that he could participate in games again. I was happy for him, but at the same time I was concerned about my own well-being. I really wanted to see my son's participation, but had no idea what to expect going into the gym. There was a Wi-Fi tower on the adjoining elementary school grounds. I knew that dealing with all the people, along with cell phones for each person, the lights, noise, and the metal bleachers would be a challenge. Now when I pulled up to the school, I would imagine someone who was on oxygen having to be without it for a few minutes in order to get situated. I had to walk in as fast as I could, rushing to get my grounding gadget plugged into an outlet so I could get it around my ankle quickly. Once I connected this grounder, it would act as a faraday cage protecting my body from feeling all electricity. I would have to sit there and literally focus on calming down. My nerves felt like many needles were sticking me all over. It was very painful. My stomach would tighten up, get hard, and I was nauseated. As I got my body to calm down, I sat very still watching my son's game. I no longer felt like the same mom. I once was able to shout, cheering my son on at his games. Feeling this sense of loss made me cry. I also felt aggravated in my body by the movement of the people getting up and down. I just told myself to keep it together and that we would get through this eventually.

Many people reading this story may want to ask why I went. Great question! I thought who was I to sit at home and not support my son? No matter what happened to me, I could never replace the lost memories if I missed watching him play his games. Brandon is my only son, these were his high school days, and I had already missed out on a lot of previous events. It really was heart wrenching for me not to be there supporting my son. Especially since I did not have the grounder gadget early on, which would have helped me deal with the electricity so I could attend more of his events. Once the game was over, I would remain seated with my mom and dad until everyone left. One amazing thing I discovered was how calm the gym becomes when everyone leaves. Then I would unplug my grounder and go quickly to the car in order to get home fast so I could get plugged back into my outlet. I was very thankful my friend, Chris, had made this improved grounder gadget. Every day I thought how grateful I was to have such a unique friend that had given me more freedom than any doctor I had gone to.

# MARCH 2011
# Certified EMF Inspector

I began to prepare for a visit from a Certified EMF Inspector coming to test my house that was recommended by the National EMF Mitigation Organization out of California. I was so thankful that a representative called me back so I could pick his brain and get some help on the east coast. I decided to tell one of my neighbors what was happening and that an EMF tester was coming to check the levels in our house as well as get a reading from the power lines. I welcomed her to come and listen herself. She did, and I am sure she was surprised by how I looked. She said they thought I was pregnant because of my weight gain but knew that was not true since they never saw a baby. She watched closely and did not say too much. I just wanted her to know in case they started having health problems. My neighbor claimed to have done some research but I never found out what they did further. I understood their dilemma. One owns a house and the power company is not going to help because they cannot admit a problem such as this exists. A home purchase is a huge investment, so coming to grips with issues stemming from power lines next to your house is something most would be in denial over thinking foolishly about any risks involved. My family was very fortunate that the power company purchased our home on the agreements they wanted the right-a-way which would have moved more power lines up to our back deck. Looking back, I do feel fortunate that this transaction took place. When the representative finished testing our home and property with all her gadgets, she said, "I do not know how you can stand up, nor have a conversation in this house. The levels are extremely way over the standards." We all must know and understand the standards set by our U.S. government are not human friendly. As time moves on, we will seek to get the standards revised. I knew I would be a part of this change. I thanked her and wrote a check paying for her services as I was relieved that this was accomplished. I had worked towards getting this large task done for a long time in case of a court hearing, but more importantly for my own knowledge. Furthermore, I had chemical analysis tests conducted on my entire property before we decided to move breaking our lease with the power company. Fortunately, Dr. Rea had a department that was capable of performing chemical testing on houses in order to reveal the toxicity levels to their occupants.

I had not had a good night's sleep at all for a very long time while dealing with this EMF sickness. It had now been two solid years of dealing with severe pain and isolation. I started to take a product that I ordered based on the information I received from the brain camp facility. I felt some relief in my nervous system with this Glutathione product. My headaches and retching eased up enough to the point I was more capable of relocating my family. I decided that I needed to find a place we could rent and that breaking our lease with the power company was inevitable. When the power company purchased our house the agreement was we would lease the house back from them for three years. However, I was not aware of what was going on the day we signed that agreement, and I was so down in my health I could not fight or take on a move of this magnitude. I had 22 years of my life along with my husband and our children's belongings in this home. Now, I was stable enough to start looking for a place to relocate us. Please realize how stressful it was for me to just go out of the house into the public. I had trouble riding in cars from exposure to all the power lines on the roads. Wi-Fi towers are everywhere, and having to drive up to intersections at stop lights was also difficult because of radiation emitting from the electricity needing to operate the large equipment. I also had difficulty with movement, noises, and odors. This was all directly related to suffering from brain damage to the parts that control those senses. This was no joking matter as crazy as all this sounds. It was my daily life battles. My sister and I went out looking at places to move to when she and her children came to visit. We saw so many places located near power lines. Apartments are too close to one another due to wireless emissions from everyday technologies, especially, for someone in my condition because I have to be away from exposure or I retch and suffer pain in my body. I finally found a house that was in the same school district so my son could complete his high school years with the friends he grew up with as they all attended the same schools. Thankfully, the house backed up to an old golf course that had no power lines nearby. I went and applied for the rental property alone and secured the place that day for us to begin the moving process. In the meantime, I knew I would have to go through 22 years of stuff and I was obviously moving slowly. My mom helped me the most with packing and moving belongings. My sister assisted some and Brandon helped as much as possible after school hours. My husband did not help at all. He still was managing the stores and had a real hard time with the changes that were taking place. Fear and denial are very paralyzing emotions and this was where we saw ourselves retiring when we initially bought the home 12 years early. This home is where we held family reunions, our children's birthday parties, our friends and family gathered for holidays here. However, as wonderful as those memories were, we all were being poisoned unknowingly and I suffered the most from the exposure. I had six weeks to get this enormous move accomplished. I physically had very little strength and was forced to work slowly. I sent a 30-day notice to the power company along with April's rent check. I started to call auctioneers to see what they would offer to just buy me out. It was like giving to Goodwill. I found one that offered to do everything as well as hold a live auction once I moved everything to the new rental house. He would move all items outside our house under his tents. April 30, 2011 was the auction date.

# APRIL 2011
## Auction

This was the month we would close on the sale of our Subway restaurant and shed one big financial headache. I had dealt with a lot of stress from my husband, Johnny. He perceived the sale as a reflection of his abilities in managing two stores as being a failure. This was so far from the truth, but I understood his thinking. I knew this restaurant was causing us to be in the red (I managed the checkbooks), and we just could not keep going in that direction as business owners. We rode together to the bank where Subway made special closing procedure arrangements for me due to my challenges. Normally, our particular location of franchises would have to go to Chapel Hill, North Carolina as that was our direct development office headquarters. I was very grateful to them for doing this transaction locally. Johnny had an attitude because of having to deal with all the sudden changes. Therefore, he only answered a question if he were asked directly. I knew we looked like two troubled individuals. However, I was just thankful the transaction was successful. I knew Johnny would eventually resolve his issues. After the signing, he made some unkind comments and I allowed him to vent his frustrations, but I refused to say anything to him. I understood he was responsible for his own words and actions and would see the truth down the road. I remained focused on our move and had to stay strong. The funds available from the sale opened the door for me to continue the quest for my healing. I was looking forward to attending the brain facility and learning more about the many variables needed to move me closer to finding the solutions to my health predicament.

It was good timing that we had until April 30<sup>th</sup> to be out of the house. I continued to do what I was able to do with our belongings. My son and I moved into the rental house alone. I got myself situated where I could function and worked like a snail on organizing our new living quarters. This was a considerable downsize as we had accumulated more than would fit or was needed for the new home. My dad and brother, Barnabas, helped with moving all our heavy items with their trucks. Still having a detective type of discretion from my former bondswoman days, I moved during the night because I did not like for people to see our possessions. Brandon and I resided in the new rental property for 30 days without Johnny. He struggled with all the changes that had to take place and was determined to stay at the poisonous house, sleeping on our former mattress set that we were going to take to the county dump. Pathetic as it sounds, this temporary living arrangement worked great for me because there was absolutely no electronics connected such as TV's, computers, or wireless Internet modems. This really helped me in getting some needful sleep since all those modern day technologies interfere with quality rest. Good sleep is needed to build our immune systems and suffering from EMF sickness makes quality sleep even more important for my body. So, Johnny's resistance to change caused him to procrastinate on packing his belongings as well as getting his sports room together in the time frame prior to the auction. He waited until the day before the auction to move some things out in his sports room. Thankfully, the auctioneer worked around my stubborn husband covering the items with blankets that Johnny had not thought about selling in our auction because he was preoccupied and overwhelmed with all the changes taking place. The auction day arrived and people showed up. As they bid on our "stuff", I could really feel relief as some pressures unloaded from my life. The house had to be cleaned and the keys turned in by the end of the day. I was feeling noticeably different taking a very high level of that Glutathione product I had ordered. I was tolerating the house and property a little better and got it all cleaned. My mother has always helped me, so she and my daughter, Brandi, were assisting in the cleaning process. Brandon videotaped a good portion of our auction and cleaning as well as took some photographs of this historical moment in our lives. We all believed this would be beneficial for the movie I wanted to make about my journey with EMF sickness. Understandably, pictures are great visuals used to document stories and solidify memories.

The following week my dad wanted me to attend a meeting that discussed different health products that was being held near his neighborhood. I agreed to go as I am open to hear about new information that may be beneficial to my health needs. I met Jenny and Wes, the couple involved with the company's product lines. I listened to all the testimonies that people were giving on how the products had helped them. Therefore, I told Jenny we would talk soon as I had to think things over, study the ingredients and formulations of their products. Four days went by. Then I called her to discuss how the water ionizer machine I was using went hand-in-hand with their products. Quality water and their meal replacement shakes would serve as a great combination, so I said I would purchase from her if she would consider buying a water ionizer machine from me. She said she would review the information on the water machine and call me back. Long story short, we both purchased from each other and I started the product line of meal replacement shakes containing protein as well as their line of other foods and supplements. By the second day, I felt different in a good way as I believed I was on to something nutritionally beneficial. Nutrition was very important for me and that I got enough nutrients into my cells every day. This product line seemed to help me feel so much better. I started to lose some weight and maintained the 30-day program the company designed with their line. This would really give me a boost before I went to the brain facility. Jenny and I became friends through our connection. It was as if we had known each other for years. Since she had purchased a water ionizer machine from me, my brother, Barnabas, and I agreed to show her how to connect it and be sure it was working properly. When we arrived at Jenny's house, I noticed that all her lights were CFL bulbs. I told her about these types of light bulbs because they really bothered me, even with me wearing my grounding device. They turned them off and used a non-CFL light for us to see in the kitchen. The amazing part to me is that Jenny had a certified green home where people would tour the house learning how to protect the earth while saving energy. Smart of her, she decided to investigate the information I shared with her about EMF/ES sickness. She was surprised to find out about the seriousness of the potential injuries that could occur from using CFL light bulbs and how sick people got from exposure to them. Their home was enormous in size; however, they changed out the entire 200 plus light bulbs in that house. Here she was a well accomplished architect just newly discovering some real truth about this so called "green movement" happening in our environment. I realized much more how our society is clueless to the effects of electricity, the smog produced by it, and the attack on our health making people sick in ways they never imagined.

# MAY 2011
## The Potato State

Well, here it is May 2011 and I'm still marching on this journey towards my healing. I have three weeks before leaving to the brain center in Idaho. Not looking forward to experiencing another airplane ride but realize something better is on the other end of sorrow and pain. Keeping hope alive and working on remaining positive is what I choose to pour my energy into; besides, I have a lot of paperwork to complete from the restaurant being sold and things being switched over. I have no room to allow myself to worry about the airplane ride. This next part of the journey was going to be five weeks long so there was a great deal of tasks I had to finalize before leaving. I had to make sure all the bills were paid for that five week period since I was not returning home until July. The brain center treatment program sessions were two weeks long and I wanted to continue my healing process by visiting my Aunt Barbara in Minnesota after leaving the brain center. I truly needed a reprieve from the everyday problems so I decided to take an extra 30 days to stay with my Aunt in hopes of improving. I just wanted to get my life back and be able to enjoy my children. I noticed that taking these products I ordered from Jenny's company made me feel like a new person. So, I sent a month's shipment to my Aunt's house so it would be there when I arrived. I continued with all the preparations needed before my departure date of May 26, 2011. May is the month of high school proms and this was Brandon's junior year. He looked really nice all dressed up in his tuxedo. Before leaving for his prom, we all went to have pictures taken by some beautiful scenery. I was not feeling well and looked sick. However, these were memories that I did not want to miss. All are a part of my story.

Since we left the old house back in April I got busy with all I had going on in preparation to leave at the end of May. I had not heard back from the power company and wanted to get the keys to the old house to them, so I called inquiring how to handle the matter. Furthermore, we had a pool in the backyard that was uncovered and full of water. I did not want a child to get in the backyard area and drown in case the gate was open. I wanted to be free from this house that had poisoned me and my family as well as get on with my life. Surprisingly, the woman said they had mailed a package to the house and that I needed to sign some papers. I told her we did not live there anymore and asked why they would have mailed the papers there. So, I asked if she could fax me the letter that was sent. When I got the fax I could not believe what I was reading in the letter. It stated that the power company would let us out of our lease if I would agree to drop all charges against them due to the proximity of the power lines and my illness. Unbelievable! I thought to myself, "What? Are you kidding me?" In addition, they wanted my signature notarized. I did have brain damage but I am not ignorant. So, I copied the letter I had previously mailed to them with my 30 day notice indicating we completely vacated the house on April 30, 2011 upon the advice of my doctors. I also sent them the keys through certified mail. I picked up the package they sent me through the mail and kept it sealed since the written statement enclosed was admissible in a court case.

I also received the report for my neurological tests done in

Dallas, Texas from Dr. Didrikensen. It stated that due to brain damage and a lack of performance, she considered me to be 100 percent disabled. This was sad to read knowing how active I was just a few years prior and now could do almost nothing. I sent this letter and report to Nick, the disability lawyer, so he would have it before I left to the brain center. Dr. Didrikensen asked if I was sure I wanted to go to the brain center since the facilitator's formal education was a doctorate's degree in nutrition. She did not want me to get taken advantage of with my illness or situation. I told her that I was confident and felt this was the right thing for me to do. I would let her know how it went when I got back. It was the night before I was to leave and I had everything done. I received a phone call from an old friend that her sister, with when I had attended school, had tried to commit suicide that day. I felt so bad for her, and we talked late into the night. I did not get much sleep. I knew she was really depressed and felt like giving up since the doctors had prescribed medicine that had backfired on her. I wanted her to find peace since she has struggled with depression. I simply had to pray for her and wait for an update when I returned home.

My dad and mom picked me up in the morning to drive me to the airport. I said goodbye to Brandon; gave him a hug, and shed some tears. I had lost a great deal of time by being away from home over the past few years. It was not fair for my son not to have his mom, but we will survive. I said goodbye to Johnny but he had a little attitude, so I was not happy with him. He had caused me extra stress by refusing to be supportive and understanding of my health problems. I understood he believed it was the end of his world since he based his success on our possessions that were, in his eyes, now a loss. Losing our poisonous home and having to sell one of our restaurants made him feel like a failure, but in fact he was simply being self-centered acting like a scared little two-year-old brat. This is what fear does to people and how it affects others close to the fear embracer. I'm the one who's sick, managing the checkbooks, trying to hold the family together, and get my own life back. However, as difficult as it has been living with someone trapped in fear, I always worked at remaining a positive role model. I knew Johnny would eventually understand and come to grips with reality. This is where the scripture says love covers a multitude of sins and I just had to forgive and continue to move forward.

I couldn't get mad. I would just have to understand how he processed life and not let it affect my healing. We got to the airport. I started to retch and cry. My mom and I were going together so we said our goodbyes to dad. He looked concerned and said I was going to be fine. He told me that he loved me and would be praying hard. We went through security and I had to be patted down. They got an attitude with me; I did not care. I had to do what was best for me. They told me that the machines would not hurt me. I knew that if any of them had radiation poisoning and electrical sickness, they would not go through those scanners either. I retched repeatedly which seemed to make them hurry. Great for me! When we got to the gate, I gave myself an antigen shot and got grounded with an outlet. My latest grounder helped this time at the airport. I was grateful that Chris was able to make this gadget for me. We boarded the plane. I retched and sat quietly until we took off. Once we got high in the sky, I started calming down. I was able to catch up on my journal and relax a little more. I was taking a glutathione precursor which made a difference. As we began to land, I started to feel sick again and retched. We got off the plane and I had to go ground with an outlet. Mom and I caught another flight to our connecting destination. It was the same thing all over again for me. Finally, we landed in Spokane, Washington and got our rental car in order to drive another three hours further into Idaho to the brain camp facility.

We finally arrived and saw where we were staying for the next couple of weeks. We both were exhausted and went to bed before dark even though there was a three-hour time zone difference for us. I felt like a limp noodle, sleeping very hard as if my body was working to repair itself from the long trip. I did not want to get up the next morning. About lunchtime, I meet Dr. Allen. She performed an ear, eye, and head test on us both.

We then went to lunch and I was able to show her my grounder as well as discuss my EMF sickness. Mom and I came a day earlier because I needed to allow myself extra time to recuperate. Dr. Allen had just returned from a conference and had no opportunity to review a book about the benefits of grounding she received prior to me flying out, nor open the package containing a grounding pad that came with the book. Therefore, she asked that I tell her all about my situation as she continued to do a pretest while videotaping me. I had very little balance. I could not jump, skip, crawl, or do much activity that would cause my head to bounce around. As she covered the test results, she concluded that my blood brain barrier was operating at a six percent level and my glutathione levels at two percent. I understood more why I had been in pain. I had no protection for my brain and cells. If someone has suffered any type of injury, born a C-section baby, or forceps and suction were used during birth, they are likely to have blood brain barrier issues. Exposure to power lines also weakens this protective barrier, especially with amalgam fillings in one's teeth.

It's amazing how we can all learn something new from others. I had to share the information I learned on CFL light bulbs with Dr. Allen and how these bulbs emit radiation in the air because of the mercury inside the glass. The next day she had changed out her CFL light bulbs. Mom and I learned about alarm points on the body, essential healing oils, and doing patterns with our arms and legs that retrain the brain responses from her. She also had a movement bed which helps with retraining the brain with movements and we started that along with light therapy. I was glad I had my grounder or I would not have been able to do the lights. I had fourteen days of this intense therapy. We discovered the trigger for my retching episodes, how to release electricity out of the body, and how to stop my nausea. We learned that brown spots on the body were free radicals and how to treat headaches with the essential healing oils. During the therapy sessions Dr. Allen also concluded that parts of my brain were functioning like an autistic brain because I was able to tell her what I was feeling during my experiences and responses to the different treatments. Since I was born with a normal functioning brain and as an adult now with this EMF damage, I know how a normal brain and an autistic brain functions. I can be a spokesperson for those who are afflicted with autism and help parents better understand what their children are experiencing. My discoveries and sufferings can in turn teach others caring for children with autism how to help lessen the tempers and the pain they have in their heads. I watched a documentary on TV while I was sick at home that covered a private school catering to autism. During the course of the documentary, the children were having a program for their parents in an auditorium, and one of the little boys said his head would hurt every time he was around the lights. The teacher said, "Oh Billy, you are okay." The mother also told him he was okay. He was holding his head while rocking back and forth. I felt his type of pain before and I just cried for this child. I knew he was hurting with similar headaches that I have experienced. I decided that I wanted to be a voice for autism sufferers down the road, which is another desire I have in helping others. I actually have another video made with Dr. Meehan on a different occasion that shows me rocking back and forth holding my head because it hurt so badly. I always worked at having proof to back my story along with science. People need the truth and I want them to be able to understand the pain I've suffered as much as possible with visuals such as videos I had recorded. Pictures speak a clearer message, I believe so we can all understand a little deeper. When I saw that little boy, named Billy, reacting to the lights the way he did, I realized the connection we had, but unfortunately the adults in his life had no idea how he felt or to know to turn off those lights or get him out of that environment. This is so troubling to watch, knowing what I know and wanting to help but cannot. Perhaps, whoever reads this book will be in a position to reach those that I may never be able to touch. I know our journey in life has purpose and meaning for others. It's always about helping others too.

# JUNE 2011
## It Is Over!

Our time had come to an end at the brain center. As I went through the entire pre-test again in order to see how much I improved, I was happily amazed! Obviously, I would need to continue exercise patterns for my brain as I was still in healing mode. I was surprised how my brain actually felt as if it were doing a complete work out. I was making new neurons in my brain. That is why it was so important that I needed to keep taking a precursor glutathione product. I was able to stop with all my old vitamin supplements and antigen shots. I was experiencing greater results with my current regimen of products. In addition, I recorded my own video with Dr. Allen explaining my situation and how I had improved in 14 days with these new treatments. The big test was coming up; seeing if I would retch when I got to the airport. Mom and I packed and left early to drive to Spokane in order to catch our departing flight. We got to Spokane and spent the night in a hotel. After getting a good four hours of sleep; we took our showers and headed to the airport. My big test had begun. We turned in the rental car and went to security to check in and board the plane. We got on the plane, making one stop in between flights, and I only retched a little. That was a major accomplishment! I had been retching for the past two and a half years. I learned how to ease the retching episodes, my nausea, and headaches with the essential oils. This knowledge alone was worth my trip. We arrived in Minnesota and stayed with my Aunt Barbara. Mom and I continued to do the patterns, and all the other things I was taught at the brain center. My mom visited relatives and friends she had not seen in a long time. We were originally from Minnesota, but had been gone for 35 years. I did see some of my mother's side of the family, like my Aunt Carolyn and Uncle Marty.   I saw ten other relatives who came to Aunt Barbara's house. I was able to put two puzzles together, which was another significant improvement since I could not previously play games that required my mind to engage, because I would have a retching episode from the movements. Every day, I was noticing more positive things happening on my journey towards healing.

As I reminisce about all the things I've learned up to this point, I am very grateful for the people who have crossed my path on this journey, thus far, who shared what they knew.   Even though, I went to many different places to work on my healing, the main thing I want others to know is that I have always done my best to remain positive and optimistic; positive and hopeful in all situations, as well as, optimistic towards the outcomes.

While at the brain center, I was introduced to some teachings by a gentleman named Dr. Art Mathias with Wellspring Ministries who taught how emotions experienced as a child literally determines why, as adults, we behave in the manner we do.   His teachings were amazing as I further learned how we react and use the pain endured as a child, to mold us for adulthood. I learned a little more about the brain's functioning processes through the experience at the brain center with Dr. Allen. I gained the knowledge of how to fix issues on contact. She played an important role at this juncture in my life which aligned with my own determination, positive attitude and conviction that I would get better. In addition, I learned that when we are in the womb, our DNA gets programmed through our mother's thoughts, feelings, and words.

This is very deep and thought provoking information. Regrettably, we are not taught this information before we become a mother. I purchased books by Dr. Art Mathias before I left the brain center, so I could read while I was at my aunt's house recovering. I was inspired to write this poem to capsule the events that had taken place in my life up to this point. The poem is titled, "It Is Over".

## IT IS OVER!

I got up late many mornings
And hurt right through the day,
I had so much to accomplish
All I could do was PRAY.

Problems just tumbled about me
And harder came each task.
"I know God will help me?' I thought
He answered, "Just ASK."

I wanted to see answers with relief
Instead, my days consisted of misery and disappointments;
I knew God would show me.
He said, "Just SEEK."

I came to God's presence,
Knocking at the door.
God gently and lovingly said,
"My child, come in."

I woke up early today with tears of Joy,
As I thanked Him for all my answers to
Suffering, Electric Magnetic Sickness, and PAIN I had endured.
Giving me the gift of this SECRET
For HEALING!

Blessings to family and friends
who helped me through this Journey!

*Beth Sturdivant*
*06/19/2011*

Now in Minnesota after leaving the brain center, I continued following the recommended products and treatment plan provided me from this latest string of doctors I had gone to for my journey to recovery. While in Minnesota, my mother, our cousin Kelli, and I went to visit my Aunt Judy and Uncle Brian in Tyler, which was about a three-hour drive. As we were driving, I used my essential oils while taking a glutathione precursor in order to be somewhat comfortable while I was riding in the car. Furthermore, I wore colored glasses to assist in calming my nerves. I had heard about color therapy in the past and now I was able to see how it worked first hand. We arrived at their house and I wanted to share what I had learned over the past few years with my uncle who had the beginnings of Alzheimer's since we both had brains that were not functioning correctly at 100% all the time. As I showed him certain elements learned, my aunt would tell him what to do, and his response was delayed. I totally related to this behavior since I had been going through the same thing for the past few years. I would like to say in this sidebar note that if you have a loved one suffering with cognitive issues, please understand that you have to exercise patience with them because their brain processes information in a delay mode. One must not expect an immediate response or answer. Yelling or correcting them over the story they tell or if they are not moving quickly enough is upsetting to the person. Seeing others do this to loved ones is upsetting to me because I would feel bad when someone corrected me. One has to ask themselves if being right really matters. Especially dealing with someone who has short term memory loss which means the information being processed in their mind is simply scrambled.

Well, we all had a great time visiting with each other and what a nice treat it was to see my relatives again. It was a pleasure to show them some things I had learned along the way to my own recovery. I certainly admired my aunt's dedication in her care with Uncle Brian's health needs that were 24 hours a day. That's a hard service for anyone to manage.

# JULY 2011
## Balloons and Flowers

Time had gone by and my 30 days of additional recovery time had come to an end. Mom and I were getting ready to fly back home. We were both missing home, as it had been five weeks since we left. I was eager to see my children and husband. I had a decent flight and manage the retching and made it back OK. I had to use my grounding gadget at the airport as well as apply the essential oils. When we came off our arriving flight and headed down the jet way, my children were waiting to greet me with a bouquet of balloons and flowers. They were really checking me out to see if I had improved. I sometimes chuckle at how observant they are as I worked to instill this character trait in them while they were little. I got home and my husband was waiting for me. He gave me a big hug and said he had really missed me. I felt like I had returned home to a much better atmosphere, which was wonderful overall. One vitally important thing that I had learned was to be actively responsible in looking out for myself; doing only what was in the best interest of my healing. My children sat with me the next day while Johnny worked at the restaurant. I showed them some of my therapy treatments. I also prayed the mother's prayer with them that I had learned to do for emotional healing from Dr. Mathias with Wellspring Ministries. My children wanted to take me out to eat even though I really did not want to go; I went anyway. We went to a Spanish restaurant and when I walked in, I saw it was a surprise party for me. Seeing all the familiar faces of friends and family made me cry because I was overwhelmed with happiness. I was able to sit in the restaurant a little longer than in the past. However, my EMF sickness is ever present and I was still not healed. Improvement means something is improving upon a worse situation, but does not mean everything is all better. Therefore, I had improved but I was nowhere near better after leaving the brain center or going through the COEM the year prior. I learned a great deal of information which cost me a lot financially. However, being an optimistic person I know there is always hope going into any situation and one must remain positive regardless of the outcome. That's me in a nutshell, so to speak. Well, we all had a great time visiting with each other at the restaurant and enjoyed our dinner along with the cupcakes we were served. Johnny and the kids were excited and hopeful with seeing my progress. I could feel that they were at ease a little more and their energy was better.

I had to catch up with all the mail and get our checkbooks up to date right away. This was not a responsibility that could be let go or overlooked for a long period of time. It was nice that the other store had been sold since that cut our bills in half, which was a relief on the checkbooks. After learning more processes from the brain center, I had established a morning routine that consisted of putting my health needs first before taking on work or tasks for the day ahead. The checkbooks and going through a pile of bills was work for me considering the circumstances I was facing. I would sit on my bed with my book "Healing Emotional Patterns with Essential Oils" and work through healing my emotions. The body has many alarm points so wherever I felt pain I would look up the alarm point and find the related emotion in order to release the pain. What I learned is that when you have a negative emotion, you can change it with positive words. I did this constantly and literally felt relief on all the tender spots surrounding my head including when I suffered headaches. I also kept doing the patterns I had learned which would stop the dizziness I had lived with since 2006. I was slowly improving.

I received a call from my friend's mom who had talked with a local judge who knew a prominent lawyer with a well known firm located in Florida. The judge was able to get me connected with the owner who agreed to review my case against the power company. I was excited to talk to him and get my papers together for him. The law firm had FedEx pick up my package. I took a picture of that too in case I needed it for future references. I was told that it would be two weeks before I would hear back from them once they received my case files. I was really hopeful since this firm was financially established to the point of pursuing such a large personal injury case like mine, and they had many years of experience with successful results. However, in the meantime, I remained level-headed considering the magnitude this type of case would have if pursued.

# AUGUST 2011
## Younger Years

Here I am, moving forward into August 2011, having to face the rest of my convictions mandated from the judge in my former DWI case in order to be eligible to have my driver's license privileges reinstated. Therefore, I called about the alcohol class that was still looming over my head in order to get my full license back by the end of August. I called for two weeks straight before finally reaching someone who could help me line up the required state class. I went to class for two weeks for a few hours each day. I made the best out of this situation. It was pretty obvious, with a cord around my leg plugged into an outlet and using essential oils on my alarm points for nausea and headaches, that something was wrong with my health. This scene gave me a great opening to explain what I was doing. We had great discussions about electric magnetic fields and how they affect the body. Furthermore, since I had a few years of research under my belt by then, I was able to explain that addictions feed a calming effect need in the nerves of a body, making one feel better. I advised them to take the time and look around at their surroundings when they felt the need for a substance. I went on to share that one problem with alcohol is that it disrupts the flow while interfering with the minerals and enzymes in the body. This is my way of interpreting my version of the knowledge I had gleaned over the course of my journey and I asked them to at least consider the insightful information for their own good. Well, I finally completed the course and was able to get my driver's license fully reinstated, which was another great relief on this long journey.

My son and I drove to Florida State University to visit the campus and meet the football coach. I was able to do this without getting terribly sick. I continued my routine therapy in the hotel room and I kept my grounder with me at all times in order to protect myself from surrounding high levels of EMF exposures. I was healing but not there completely. We had a great time being together. Especially considering all the time I had been away from home. We toured the campus, took pictures, met with the coaches, and Brandon completed a minicamp the college offered

offered football players. After three days, we returned home so Brandon could have his high school senior portraits taken. I was able to go and watch. I took pictures of him with all his buddies and videotaped those boys being silly. This was a lot of fun. I was thankful that my children always enjoyed having us around and were never embarrassed by us.

High School was starting and Brandon would be playing football again after being out a season due to his injury. I was excited for him and happy that I would be able to attend some of his games. When I went to the games, I had to sit away from the field lights and use my essential oils at least fifteen times during each game in order to help control my symptoms. The season went on and October 2011 was coming, meaning that I was turning another year older, 41. I was 38 when I got really sick with all these overwhelming symptoms of EMF toxicity (radiation poisoning).

I heard back from the law firm from Florida, and they decided not to take on my case against the power company. They stated the usual legal lingo, "You have a great case, but we cannot take it on right now". They wished me the best of luck and for my health to get better. Momentarily, I was annoyed over the fact that I could not find one law firm in this country to help a serious problem. I could not give up and would simply have to try another route to get the word out as I was on a mission towards my own healing. Therefore, I had prayed again about what I should do and had received peace in not pursuing the lawsuit any longer. A lawsuit was too much stress to endure and stress interrupts a person's ability to heal. I decided to approach my healing through educating and assisting in improving the health of other EMF sickness prisoners. I have never been one who promotes fear or paranoia. I simply want to find the cure to my situation so I can heal and in return help others in the process.

I must take you back now to my past for the last two decades so you can see what type of person I was before life took a spin placing me here on this journey today.

## My last 20 years 1991- 2011:

I am a mom before anything and I love my children. I am saved and believe in the blood of the Lord Jesus Christ as my personal Savior. I love to pray and to help others. I believe in respecting my parents. At a young age, I learned we could agree to disagree respectfully. I am a super multi-task type of person. I was the incredible perfectionist who would paint her baseboards twice a year, vacuum her furniture once a week, keep all drawers neat, and clean the edges weekly. I kept a well-landscaped yard, pool and a clean car. My husband's words about being married to me were: he had married a plumber, painter, and landscaper who could fix anything as well as prepare meals quickly without making a mess. My restaurants were so clean one could literally eat in the bathroom. I am the middle child of five, with two older brothers, one younger brother and sister. My role has always been to keep balance.   I am a problem solver, not a problem maker. I love to be busy working and to have a lot of things going on at one time.  I have been an entrepreneur my entire life. I started selling candy to classmates in ninth grade - Blow Pops to be exact. I considered this a serious job that could make money for me while I

I am having
all pregnancy
symptoms.
Mother and I
watching my
son's game.
February 2009

In my thoughts I felt like a big failure and worthless because I was gaining so much weight. I wanted to commit suicide. March 2009

I tried to calm myself down from this mess that I really did not know anything about my situation other than the fact those power lines behind my house were very dangerous. April 2009

Received devastating information about my health after living in this house for nine years and as part of the selling agreement we were to rent for three more years. This was tough considering the house was poisoning me and my family. April 2009

I was informed by Dr. Rea that power lines and chlorine pools are a lethal combination on the human. Our families and friends would come over every summer enjoying the pool. None of us knew the danger either. Fall 2009

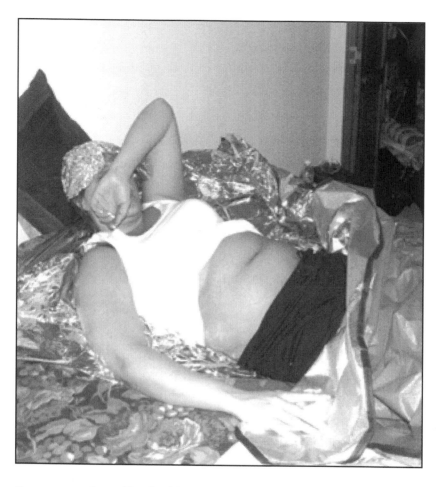

I was constantly swelling by this time with no hope in sight that I would stop.
It was out of control. I could not deal with any kind of lights. May 2009

I was extremely miserable this day, only to find out my parents had installed a Deck phone for home use. My face has numerous brown spots by now which I learned later on was an overload of free radicals in the body. I was tolerating a little light exposure. July 2010

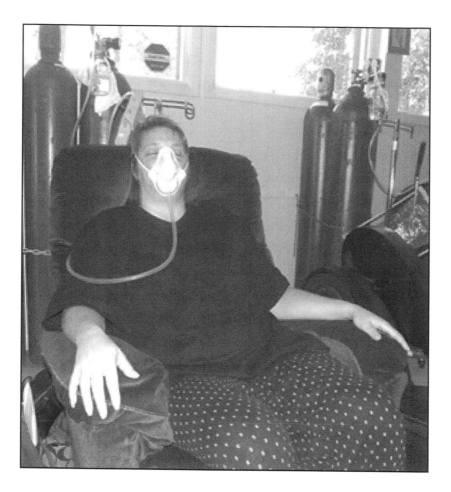

Every day I would have two hours of oxygen treatments using a porcelain mask in order to help the brain. September 2010

I could not believe how I
looked having my 40th birthday
in a bio detox center!
October 2010

I was happy and excited about
going to see Dr. Rea in hopes
of healing my situation.
November 2010

I had numerous allergy tests done at Dr. Rea's to find out what heavy metals
were in my body. He created an antigen shot  that I would take when I was
having a retching reaction. November 2010

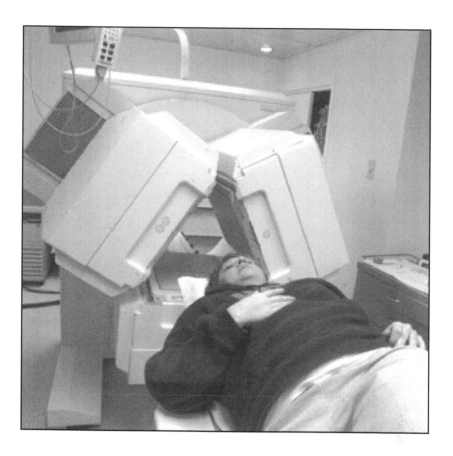

I had to undergo a nuclear brain scan in order to show my brain damage.

Another Christmas at my parents arrived, but this time we had no tree and no decorations at our house. Everyone was simply worn out from the past year. My situation had taken a toll on the entire family. December 2010

When I walked past the kitchen, I saw the power company representative in my backyard beyond the pool during the snow storm. He was trying to get a reading with some equipment. I quickly snapped a picture through the window for my own personal evidence that they came to the house. January 2011

The last night I had to sleep under the power lines.
The next day I was moving to another rental house. March 2011

Twenty-two years of our belongings auctioned off. A bittersweet day my family had to endure, but happy to put this chapter behind us. April 2011

The auction day came and people showed up bidding on our items. I could really feel the relief from some pressures unloading in my life. April 2011

Me and my daughter after 30 days of not sleeping at our house
that poisoned us. April 2011

The law firm had FedEx pick up my case file that I put together. Everything was documented thoroughly. I took a picture in case I needed it for future references. This was the last time I would present my documentations. You will have to read the book for the answer!

This was a major accomplishment for me—standing on this football field exactly one year after my 40th birthday. The metal bleachers and the high potent lights inside the stadium were extremely challenging for me. When it was physically possible for me to attend my son's games, I had to stand barefoot to ground and practice "mind over matter" so I could watch him play. On this night, I was determined to attend my son 's last homecoming ceremony. October 2011

Imagine walking into a restaurant being allergic to electricity. I had some remedies that helped get me through the moments. Going out to functions like this set me back in my health situation. However, I would just have to weigh the odds out of my role as a mother. Most moms will do almost anything for their children's benefit. I am no different. December 2011

was in class. I advanced to selling a variety of candy in high
school. I sold 200 dollars worth of candy a day and I never ate
my profit. My parents would take me to Pace Warehouse twice a
week during 1987/1988. I carried my books in one bag and 50
pounds of candy in my other book bag, which held the box of
candy. I would stock my box, sell between classes, and stay on
point while walking to my next class so I would not be late. I was
known for selling candy and because I took it seriously, I eventually was able to buy a brand-new car, a Nissan Sentra, plus pay for
my own insurance. I needed the car in order to be able to drive
across town to another high school that offered cosmetology
classes. I never cared to do hair but wanted to own a salon, which
I did later on in life. In the school district I attended, they worked
in conjunction with the state in order for students to acquire
credit hours towards owning a cosmetologist license; something I
wanted to earn at the time. I graduated high school and did wait-
ressing for a bit. I moved into my own apartment with my high-
school sweetheart in 1989. I gave birth to our daughter in August
of 1990. That same year, I lost my great-aunt Jeanne, whom I
loved dearly and would visit once a year during high school for
two weeks at a time. We really had a great bond between us. When
she passed, she left me a small amount of money that I later used
towards my second successful entrepreneurship business.

I decided to start my own bail bonding company. I was the
youngest bail bondsperson in the state to start his or her own
company without working for someone else. A local news station
did a special on me and my company. The reporter followed me
around for eight hours, observing me doing a bond skip; taking
someone to jail, and being a mom with this type of job. They

aired a two-minute segment. I was 21 years old at the time. I had met a lady, named Omega, who was also a bail bondswoman and she became a great lifelong friend. Omega mentored me early on if I needed help or had any question that needed an answer. We became good friends and pursued many bond skips together. We have so many stories we could tell. One is my very first bond skip. I was a little nervous but understood what I had to do. I called Omega and told her the situation. We had to go to this hotel that was located in one of the worst known drug areas at that time. The absconder's girlfriend had called telling me he was getting ready to leave and that I needed to come get him. A large crowd was hanging around outside the hotel and it was obvious that we did not belong. I knew that if we went in there, we would have to take the man out in handcuffs and my reputation as a bail bondswoman was on the line. We entered the room and I said, "I am here to take you to jail". He was a stocky and very muscular young guy. He said, "I am not going anywhere". I said, "I'm not leaving without you, so we can go nice or we can go ugly". He crossed his arms and lay back on the bed very stiff until he was shown what we had to make him move. Then he became serious. He immediately got up and said, "Yes ma'am". I was able to cuff him, walk him out, place him in the car and take him to jail. I really had to laugh inside because I was just as nervous as he, but did not show it.

I continued bonding for a few more years, but decided I would need to find something else to do. Omega and I had a lot of good times working together even in stressful situations. However, I would not be bail bonding the rest of my life.

It was now 1993 and I decided on owning a Subway franchise. I proceeded and asked my parents if they would like to be partners, to which they agreed. We sent our paperwork in on August 28, 1993. This process continued until January 1994, when mom and I were granted a franchise. I was now pregnant with my son and would include him into the equation. He was born in March of 1994. My husband's father passed away in August of 1994. The day my mother and I left for the mandatory Subway franchisees training, her mother died. Thankfully, her family decided to postpone the funeral until my mother could go to Minnesota. They felt that this (my mother and I attending the required training for Subway ownership) is what their mom would have wanted for us. Mom and I finished and arrived back to our home. My mother flew up to Minnesota in order to proceed with her mother's funeral taking my brother Barak. Barak had taken on all of my bail bonding work while I was gone. Johnny, my husband, took care of our children. Brandi was in kindergarten and Brandon was seven months old. I had only been home from the Subway training for five weeks when Johnny's mother went into the hospital. She passed away in December 1994. My dad's great aunt also passed away. This was a hard time for me too since I loved my mother-in-law. We were like two peas in a pod and she was so excited about the Subway venture but never saw it come to fruition.

January 1995 came and the construction on my first Subway restaurant started. We opened in April of 1995. My husband and mom worked along with me for 12 hour days on a rotation basis. I would usually take my mother's shift and have her watch our children. This was better all the way around, except for working with my spouse more as it's tough on a relationship to live and work together. This schedule continued, along with the bail bonding and being a mom of two small children. In 1997, I decided to open Marie's Salon of Elegance. I wanted a self-run hair salon. I worked hard to remodel the building and getting the business up and running along with my parents on this venture. We had ten styling booths, three nail booths, a tanning room, and massage room. I sold many products and collected booth rentals once a week. It was a great experience and I loved it. I sold the business within two years and decided to get rental duplexes with my dad and brother. We did that and my goal was to own the street of duplexes. After getting five units, I decided to sell. They sold in less than twenty-four hours, which took the realtor by surprise.

My husband and I finally officially tied the knot in a little town in SC. My children were so excited about us getting married, but very sad we were leaving them behind for our honeymoon. My parents had a very nice surprise reception for us when we got back. My children were so excited to help my mother plan the event and were able to keep it a secret.

A few months went by and we decided to move in to a new home in 2000. We wanted to be moved in before Brandon started his school year and Brandi would be going in the sixth grade. We moved into a brand-new house that would eventually change our lives forever. Once in our new home, I immediately started working in the yard and decorating the inside. Our goal was to have a fun yard and a kid-friendly house. I continued to bail bond, operate the Subway, be a mom, and do the landscaping around our home property. We installed an in-ground pool, cut down twenty-five trees, built a playhouse that was 12 by 12 with a porch on stilts, had horseshoes below, and build a half NBA court with concrete. I was preparing for a big 60th birthday party surprise for my dad to be held on our new property with guests coming from all over the country as my father had family and friends who lived in other states. I was preparing my home and yard for this big outdoor event. In 2002, my husband and I bought my parents' portion of the Subway franchise restaurant. In August 2003, my mom and I bought the Majestic Coffee Shop located in the same strip mall with the Subway. I continued there as I started building a second Subway about three miles away. I opened another Subway restaurant in September of 2004. My plate was a little full, but good. Johnny was running the first Subway and I was running the new one. My mom and I were sharing time at the coffee shop until we sold it in May of 2005. The town's mayor purchased the store from us. This was the first investment I almost lost out on financially, but didn't.

After the sale of the coffee shop I was down to just owning two Subway restaurants, and being a mom. Now, my daughter was turning sixteen with a big party planned for her. She was also a senior in high school this year. Brandon was an eighth grader. This is about the time I noticed the severe anxiety attacks I was under and feeling unexplainable craziness in my body. I started having dizzy spells but assumed it was due to stress from life's pressures. This went on for about two years. Then my back went out really bad. It was so dramatic that I could not get up from the floor for 60 days. It had never happened like this before and thankfully I had a couple of friends there along with my daughter to help me. My two child hood friends, Caroline and Missy, were there, along with Brandi's good friend, Jeff, when this happened. Caroline decided to stay with me for a month before school started. I had an unforgettable episode one day when I went to the bathroom, and suddenly broke into a sweat like I had been in a steam room. I honestly felt like I was going to die. I was in the most excruciating pain, alone in the bathroom with the door closed, and I could not speak above a whisper. This was so overwhelming that I sincerely wanted to die and nothing else mattered. My little dog, BJ, was going crazy scratching at the door and no one could hear me asking for help. Brandi was shut up in another room doing a procedure on her back for psoriasis and could not move. Finally, she yelled for Caroline to see what BJ was scratching the door about and she thought she had heard me asking for help. Caroline was on the other side of the house so she came to check on me and was frightened by how I looked and I could barely speak to her. Caroline said, "Oh my! What's going on with you?" She went to get a fan and a cold washcloth so I could cool down. I finally came to and got myself back together. I went back to our den's floor where I had been staying for the past few weeks. I felt something inside me that day that acted like a switch had been turned on/off, and then I saw my stomach. I thought it was swollen due to me lying around having no activity.

I continued to try and heal from all this drama with my back so I could get back to work. Brandon was starting high school in the ninth grade. I was able to go to his games although I was moving slowly due to my back being out similar to bulging disc or something along those lines. I started feeling nauseous and having headaches. In October of 2008, the power company came knocking at the door asking to purchase our house. We came up with a temporary leasing agreement in order to lock in the transaction. I decorated the house really nice for Christmas 2008, as this would be our last one in the home, so I thought. The Christmas decorations and Christmas morning all turned out great for the family. However, I was in a state of high anxiety, nervousness, and feeling constantly nauseated. I was noticing that my breasts were leaking as if I were pregnant and lactating. I was vomiting without warning, which would come on fast and unexpectedly.

The 2008/2009 holidays were over and Brandi was headed back to college. She wanted me to see a doctor before she left so we could see what was going on with me. I did go to the emergency room since it would be some weeks before I could get to a doctor. I decided I would do a pregnancy test, which came back negative. I was worried because I knew something was not right. In my mind, I thought it was just too early to detect. So I would wait for my appointment with the doctor to see what they would be able to tell me. My first appointment with the OB GYN came and I was told I was not pregnant, but had a large cyst on my right ovary. I asked the physician if that would cause a woman to feel pregnant with these symptoms I was experiencing including my breast leaking? No, it did not, but they would schedule another appointment to watch the cyst. Then the DWI happened, which is recorded at the beginning of this book. So, now you have a better picture of who I am.

# OCTOBER 2011
## Soul Search

It was now my 41st Birthday. I had been living with this electric magnetic sickness for a while. I was healing, so I thought, but had not quite figured out what more was needed. It seemed that once I acquired all the knowledge I could have needed to heal, someone would show up with another product that would help me improve but never truly heal. I met a lot of new people from other companies and a formulator who was able to further expound on ingredients needed for the brain to function more productively. I just kept building on the knowledge like being in a long collegiate program. I did a lot of journaling and anytime you are in a chronic fight of sickness, the most important thing you can do for yourself is to write about your day. This evidence gives proof of how far you have really come. For a few years, I had a bad day every day. Then I would have a good day here and there. Now here in October 2011, I was having mostly good days and a few bad days. I had to soul search and get a reality check at times. I could really see my accomplishments and saw that staying positive was the right way to follow.

# NOVEMBER 2011
## Update

I decided it was time to schedule an appointment with Dr. Rea and give him my update. I emailed all the doctors I had seen updating them on my progress since leaving the brain center. I told them that I was still healing and was not totally there, but believed I was getting closer. My heart's desire was to find help for the EMF Community as I found my own healing.

# DECEMBER 2011
## College Grad

The holidays were approaching again and we were living in a different rental house. Brandi was graduating from college and I would be at the ceremony with Johnny, my mom, dad, Brandon, my brother Barnabas, family friends Jeff and Duran. Going to a facility of that size was a little harder for me to endure with my EMF sickness. Imagine how many people, cell phones, lights, and electronics would be under one roof. My body had become a real testing meter and recognized high levels of EMF in the air and surroundings very quickly. I always went with a positive attitude that I would be just fine. My body would sometimes react and respond severely. I always carried my essential oils and made sure I wore my diodes for additional support in preventing horrible retching episodes. Those are very painful and annoying to go through. This was a great celebration and a joyous moment in time for me and my husband to have the first child on both sides of the family to graduate from a well respected college with a full four-year bachelor's degree. We were very proud of her and all of us went out to a Brazilian restaurant afterwards to celebrate her accomplishment. I did as well as I could in the restaurant considering my circumstances. Brandi would be moving back home until she decided what her next steps would be. I would enjoy the time she stayed with us. We did not have a tree or decorations for Christmas but we were a family doing our usual holiday breakfast together. We had all been through a great ordeal. Soon we would be preparing for Brandon's Senior Prom and high school graduation in a few months.

Brandi and I went around and had many meetings. We were educating people about EMF and the importance of an antioxidant glutathione precursor plus the essential oils to protect the brain and stop pain. We really met a lot of people and I spent a lot of time counseling. I had spent a lot of time over the years, watching a little TV when I could, but seeing all of these problems people were having. I would have loved to explain to pregnant women who were celebrities that the laptops they were holding next to their bellies were affecting their unborn babies. I watched several reality shows and saw one that had a very large family with the mother using a computer to teach the infant how to talk. I was disturbed, as would be expected, and wanted to tell them to get that computer away from your newborn baby's head immediately. She has a blood brain barrier compromise that actually hurts the head and you all need to get some EMF diodes. The ones I was currently using seemed to help me the most and also helped in getting the glutathione levels back up in the body. I need to go on the Oprah show with my best friend, Caroline, and tell the world! She and I have always wanted to share our knowledge with her because we both grew up watching the show, caring about her well-being. One thing for sure about being overweight is that being exposed to high levels of EMF causes the cortisol levels to rise and the cells in the body are constantly crying out for a glutathione protector.

I am in no way a doctor, but I have studied my situation exten-
sively and experienced the extreme pain that comes from EMF
exposure. We are all affected one way or another with Electric
Magnetic Fields (EMF). If you are reading this book, you are getting first-hand knowledge from real-life experiences. You must
avoid fear, and take action immediately. For our children, today's
cigarette is the cell phone. In use, it causes anxiety, depression,
headaches, mood swings, lack of concentration, constant interfer-
ence with brain and nerves, just to name a few side effects.

# 2012

# FEBRUARY 2012
## Hearing

I had my disability hearing this month which was sixteen months after filing. I met my attorney at the hearing to go in front of a judge. I was very sick while in the waiting room and actually retched repeatedly. It was from all of the people and the electricity surrounding us, along with me being nervous. My lawyer was going over what kind of questions they were going to ask me. Brandi was with me, but they did not let her in the hearing room. Only my lawyer and I were allowed in the conference room. I sat at the end of the table and could feel the EMF immediately. The judge stated to me that in no way was she making light of my situation, but she did not understand and was not familiar with electric magnetic sickness. I was asked many questions. At the end, my lawyer gave his closing arguments and the judge asked the caseworker if there was a job that basically had no electricity. She answered no. I was dismissed and she told me that I would get an answer by mail. My lawyer told me he thought it went great and that I had answered my questions well. He would let me know as soon as he heard something and told me to take care of myself. As Brandi and I left, I felt a strong sense of peace. I had no control over the outcome. All I could say was that I had been very sick and sensitive to electricity. In addition, this situation had not allowed me to work. I would wait for my disposition, but I felt like this case was behind me and further helped those of us in the EMF community.

# MARCH 2012
## I Won

Well, I decided that I would do another round at the brain center and I was able to go with my new friends Denise, Dr. Carol, and my brother, Barnabas. I was going to help out some in exchange for treatments for myself along with my brother, Barnabas. We had the same flights and rental car as the year before when my mother and I went to the center. I did well at the airport and on the plane. We arrived and settled down. I was dealing with some pain and feeling sick. It was a long day of traveling. Dr. Allen gave me a different gadget to put on my body in order to calm me down, which it did right away. We ate dinner, unpacked, and turned in for the night. With all of my determination, I was determined to figure this out. I was not perfect yet, but I was at least 70% or more better. We were there for 16 days and enjoyed each other's company and learning experiences. I know I was improving since my last visit because my head did not hurt as badly as the year prior. I had everything done including the checkbooks for our Subway restaurant and all the bills ready for Johnny. Sixteen days is much easier for a business manager to manage being away than the five weeks I was away the previous year. My family was once again happy to see me return. I had improved a little more and we all were ready for me to be better. Improving is moving forward.

Upon my return home, I had a voicemail from my disability attorney stating, "Congratulations. You won your case. Call me when you get back." I was so excited and felt like I had accomplished what I had set out to do when I first got sick. This win was a landmark case according to my attorney. He said, "Do you realize I am probably one of the only lawyers in the country that knows about EMF sickness? Do you understand that you won this case because you gave me so much research and your Dr. Rea was able to make me fully understand what this sickness does to the body and how it disables a person?" He thanked me for trusting him to handle my disability case. I thanked him, hung up the phone, and just cried with relief.

It was like having a part of my healing project completed. I genuinely helped the EMF community by following through on this disability case and now winning my fight. I did not plan on needing the funds long. However, when money is extremely tight and the financial cost for me to do what I needed is high, every little bit I receive helped. I called Brandi, Brandon, my husband, and Mom to share the good news. I believe this win finally registered with Johnny. He understood and accepted that what I had gone through was real.

I understand why Johnny thinks and processes information the way he does. He views being sick as if one is going to die like his parents did when they got sick. His mother was my age when she was diagnosed with kidney failure and passed away at the age of 43. We were almost her age now and Johnny would be forty-three this year. It was time for me to start to write my story for a book. This was big news winning this disability case. I, and the EMF community, had to be heard.

# APRIL 2012
## Well Springs

Brandi, Caroline, Ashley, and I went to a Wellspring Ministry Conference in Georgia to hear Dr. Art Mathias' seminar on the connection between our emotions and healing the body. I knew we would all benefit together and I wanted to meet Dr. Art Mathias in person. He really inspired me a year earlier when I heard his CD's on Freedom. We met with 14 more people and all of us lodged at a beautiful cabin that belonged to a friend of Denise's. It was really nice and three miles back in the woods off the road. I was able to share my EMF sickness journey with a large group the night we arrived for the conference.

I got to meet Dr. Art Mathias, which was great. I was very impressed with his teachings, even more so in person. We heard his testimony of how he was sick and suffered in tremendous pain. He was given a year and a half to live by the doctor's prognosis with his disease. He has done extensive research and thoroughly studied how releasing emotions will release chemicals in the body, which will help the body in healing. He and I prayed together for me. I asked him why I was still swollen and having problems with electricity. He said I had a lot of fear. I thought "No, I do not." I was not even concerned about electricity like one would think since it constantly bothered me. However, I started to ponder what he told me and realized I did have some fearful issues. Such as, what if something happened to Johnny, what would happen to our restaurant? Then I remembered our children and thought what about the effects on them being raised under power lines and would a side effect from the exposure interfere with their ability to have normal children, if any children at all? What about my parents and their hurts from family? The list of fearful thoughts went on in my mind. Dr. Mathias told me to follow his instructions from his prayer book starting with the two simple prayers, one for forgiveness and one for repentance. I must pray for each item and release all hurt in my life as well as repent for any wrong I had done. Dr. Art Mathias had prayed to release the EMF bothering me and to heal my cells the way God designed them to be. I was released that night.

When you put science and scripture together in one book, you cannot go wrong. My daughter and friends were helped a lot. We all left with a great gift of more knowledge and peace. When I got home, Brandi and Brandon did the prayers with each other while Johnny did the father's prayer over our children. My mother started using the prayers from the book as therapy for her own life. I counseled people teaching them about the alarm points, essential oils, and praying the prayers with them because it does work. All his information ties into a book that was the catalyst to my understanding the power of emotions called, "Feelings Buried Alive Never Die" by author Karol K. Truman. On page 135 it reads, "We cannot learn and grow when we blame others; we relinquish our agency, our will, our freedom of choice, and our personal power". This made me realize I will be responsible for the effects of the power lines on my health and had to release as well as cancel the power company's debt to me. (Page 218) "Remember, there are only two main feelings, LOVE and FEAR. Acronym for FEAR is: False Evidence Appearing Real. Do not accept another person's emotions and problems. They are like weeds. They never go away unless you cleanse and destroy the roots". This was a great book from front to back, a must read.

# MAY 2012
## Prom

Brandon's prom was this weekend. We would go take pictures, and see him off. He looked so nice. As we got ready to go, my ear started to hurt. We were going near our old house that had poisoned me. I knew my body and brain had memory. I was doing my prayers and trying to get rid of this ear pain. I started to feel really bad and my ear was getting worse. I hung in there and took pictures. We got done, saw him off and my ear really started to hurt worse. It was getting worse by the minute. I got home and got on the bed. I was in a lot of pain. Brandi was doing the alarm points and reflexology points on my feet. I was using the oils and doing the prayers. The reason why I am writing this is because you, the reader, needs to know that sometimes it will seem like this procedure is not working. However, the problem for me was I needed to let someone know I could not come and help them with another project. Therefore, as I put off being truthful with this other person, my earache was an emotional trigger point. The time was quickly approaching that I was supposed to give an answer as to whether or not I would help. It was bothering me that I was going to leave and be gone for two weeks. My sub-conscious feared leaving again and this was causing my body to create pain. When I made the phone call, letting the person know that I could not help them, immediately the pain stopped. By now, I was becoming more in tune with my body and emotions. My self-awareness had grown to a much deeper level.

# JUNE 2012
# High School Grad

We continued to be a family and prepare for Brandon's high school graduation. We were very proud of him and I really knew that Brandon had gone through more than anyone. He is the youngest and Brandi was away at college most of the time. I was living from place to place and Brandon witnessed me being sick so many days. I am so thankful he kept it together and was able to maintain his grades until he finished high school. Brandon graduating was another great day of accomplishment for me and Johnny. Both our children finished high school with no problems.

Summer came and went as I continued to take the glutathione precursor, use the essential oils on my alarm points and prayed the prayers I learned. I did have a little more in my routine, along with a great diet and drinking alkaline water. I was reading another book from Dr. Art Mathias called, "The Continuing Works of Christ Exposing Unbelief", and on page 107 I read about "Detrimental Effects of Excessive Cortisol". As I was reading this section, I had a major breakthrough. I realized that when our bodies are in chronic stress, the fight or flight mode happens over and over again. This produces nearly 240% more cortisol, which causes weight gain and damages our organs over time. I finally grasped the answers for all my weight gain and now understood why I swelled more when I went out in public. The high levels of electricity were a stress inducer on my body. This is also true if a person is causing one stress. The flight and fight mode is continually wreaking havoc on the body and organs. This was a huge light bulb moment for me.

# AUGUST 2012
## Totally Isolated

This month I have decided to be totally isolated and choosing not to talk on the phone with anyone who has conflicting energy that causes me stress. In other words, I am avoiding negative people. After learning about what that has done to me, I am avoiding negativity like a plague. I'm 42 years old and I have plenty of prayers to work with in releasing past emotions. I have grown leaps and bounds by finding what helps EMF sickness/sensitivities. These prayers of forgiveness and repentance stand against science because I am the one who can attest to the after affects as well as have lived through the pain of EMF sickness.

I did accomplish the disability case in light of my sickness. I desire to work in conjunction with legislators and insurance codes to accommodate people suffering with EMF sickness and sensitivities through my charity, The Academy For Environmental Sickness. I have traveled down this road considering how I could help others with answers and provide support. I appreciated the first book I read about EMF sickness more so after I gained the knowledge and insight from my own personal journey. I wanted to tell my story here so many may understand how EMF sickness affects the entire family, not just the individual.

I pray for anyone reading my story who can connect or relate, receive hope and help. Furthermore, believe in yourself and know we are all-important. We deserve this acknowledgement and it's one step at a time to the finish line.

# NOVEMBER 2012
## Moringa

I was sent a product from another childhood friend, Donna. I saw her weight loss picture on Facebook and she just looked so much better. The last time I saw her was 2010 when she drove me to the airport in Dallas, Texas. I asked her what she was doing. She sent me some video links to watch. Two weeks went by and I had been thinking, "What does Miss Donna know about health?" Her family did not eat healthy or even care about nutrition since we were children. Something spoke to me encouraging me to watch the links she emailed me to just see if this information would be helpful. I looked at a Discovery Channel Documentary on the Moringa Tree and all its natural properties. I was really impressed with what I was seeing and hearing. The nutrition value in this tree is amazingly higher than any other tree in the world. This is what you call whole food plant based nutrition. I was thinking if this really was what the documentary said, I would get better and this could be the missing piece to my puzzle of healing. I was very excited to get started. The next morning would be the 30th. I drank my first pack of this product and was amazed with the taste as well as the feeling I was getting with this drink. I was also thinking I was going to have to have more than one pack a day.

Therefore, I got Donna on the phone and she wanted to send me a book to read about this miracle tree plant. I was amazed when I read about the importance of each amino acid and how the right amounts have to be consumed in the correct ratios in order to work in our bodies. Furthermore, the book explained the role of each amino acid and how each one interacted with the vitamins and minerals processing in the body's digestion. I soon discovered that I was taking numerous products containing one of each amino acids represented in this one plant called the miracle tree.

Fortunately, Donna was able to get Dr. Fisher and me on the phone together to discuss a few things about this product. I had just finished reading his book, "Magic Myth and Miracle." Since I had already researched information on my own, my only question was if I took more than the suggested amounts would I be wasting the product? Being that I was not his patient, Dr. Fisher, stated his disclaimer that he was not my doctor and he was not diagnosing, treating, curing, or mitigating anything. Clearly, I understood his disclaimer and agreed. However, he did tell me that if I could take more, the faster I would get back to functioning more normal. Our bodies must have so many protein strands to be able to function right. What this means, in laymen's terms, is the more complete amino acids one consumes along with the nine essentials amino acids, the more protein strands the body can produce. I was totally getting this now and already knew that nutrition getting to a cellular level was most important. We talked for about 20 minutes discussing EMF sickness issues. Dr. Fisher was very familiar with Electric Magnetic Sickness. So, I suggested that maybe one day we could compare notes on my side of pain and suffering. He responded with the usual comment as if my notes were not worth his time because he apparently assumed he was an authority on EMF sickness. However, I must tell all scientists as well as doctors, you can study all you want and understand how EMF sickness affects the human body, but you cannot relate to the pain or suffering until you physically endure the same pain any of us have endured from high levels of EMF exposure.

This brings me back to how fortunate I am to have the understanding family I have. Well, at this point, all I can say is I was so excited with this new information and ready to experience the healing power of this product. I decided to order more of this product and cancel all my other previous products that were set up on autoship. If I was not going to consume a product, I was not going to sell it or those product lines. I do have a moral obligation as a leader and I felt that I must share what I am doing with other people who were on the other products about my change in plans and products.

Therefore, after being on the product for two weeks, I started making phone calls to family, friends, and former doctors that treated me telling them all about this new one miraculous product.

I am so challenged and looking forward to turning my body around from this journey I have been through. I know as soon as I can get this extra fifty pounds of fluid out of me, my story will inspire others around the world. I have spent several years in isolation and been denied an active lifestyle, especially for a 42-year-old woman in our society.

I have a lot to catch up on and still accomplish. I continue to believe I will achieve. I do picture what I want and I speak it. I believe we either curse ourselves by our thoughts and words or we encourage. I always encourage and speak what I want into existence. I know throughout this book I have spoken about emotions, so I will say again be careful for what you think, causing you to feel, which causes you to speak. This is the domino effect, powerful actions but true.

# 2013

# 2013
# Got Genius

Been on this product now a good month and amazed at how well I am doing. Therefore, I decided to venture out of my house. My son and I went to our local Sam's Club Warehouse. Normally, in the past, I had to leave quickly, but this time I did not get that sick feeling with my nerves. I was able to go to the grocery store another day without the nausea and nerve sensation running through my body. I was even more surprised when I went with my son to the mall. I did carry my drink with me and was able to purchase a couple of clothing items for my son before leaving. This was rewarding to me because I have not been able to do this for many years. I was unable to hang out long, but this little bit of time with him at the mall was wonderful.

I have now passed day 90 on this product. I am expecting big results to start moving faster now. Especially now with the help of this new amazing machine I am using called the Photon Genius manufactured by The Ed Skilling Institute out of Arizona www.edskilling.com. If you are sick, you may want to own a machine or find one that you can pay to use in your area. I have had my second 35 minute session. I am experiencing a change in my nervous system with not being as sensitive. However, I am not sweating. I am using the machine on Tuesdays and Fridays, resting in between to help my body and drinking my special drink.

In order to heal, you must have three parts, emotional release, nutrition that can get to the cellular level, and physical. I can say I have worked on emotional release so much, I feel no weight from any part of my past life. I am getting superb nutrition in my body like never before, which prepared me for this physical part, the Photon Genius machine.

The following is a wrap up of all I've done up to the point of publishing this book:

Started with drinking high alkaline water from an ionized water machine, attended 12 weeks at an environmental bio detox-ification center, attended a 16 day brain camp facility with two ses-sion rounds, took a load of supplements, worked with several product lines, went through the Wellspring Ministry for emotional healing, currently on a special blend of drink mix containing the Moringa Oleifera Tree, learned and got certified with my daugh-ter as a Riki Master Energy course that uses energy in our hands to promote healing, and now I own and use daily a Photon Genius Machine. Furthermore, I have spent thousands of dollars over the years on my journey towards my healing. I am here to help others who are interested in cutting through to the chase because my past has not been lived in vain. My daughter and I have created a foundation out of my pain and sickness to help others through-out the world called The Academy for Environmental Sickness. I have met numerous people along my path and ran into some old friends who have been a part of God's plan for me on this jour-ney. I am very grateful to many people whom I have listed on the Acknowledgement page of this book and anyone who has been a part of my healing journey.

My message to everyone reading this is do not judge others, and remember you may never know why you cross paths with people who are different, or who will be the one from your past that could help save your life today. So don't jump to rash judgments of others. God is so good to me, unconditionally good. I thank him for this opportunity and want all who read my story to know I give Him the credit for my healing journey, first. I have had many dark days and all I had to do was pray. He gave me peace immediately and would help me come up with a needed plan to solve the situations I faced daily. I did not want to pass judgment on any one I mentioned. I was honest as I could be in hope to help another couple realize that you must put your fears and lack of knowledge of the sickness you are facing aside, and just be supportive of your spouse or partner NO MATTER WHAT.

I have figured out the remedy for gaining my life back. I will continue this journey of healing to cross the finish line. I am excited because God really has given me a better life this time around. I have learned to not let one person or thing steal my peace or cause me stress. I do believe that whatever is going to happen will happen. I pray that my book will provide the hope that you have searched for and the answers for the family member who is doubtful. Please feel free to contact my organization through our website at www.backyardsecretexposed.com for further information of products, donations needed, or services available to help those in need. I must say in everything give thanks and to lean on the Lord, not yourself.

I only wish much love and happiness for you all!

# TESTIMONIALS
# AND LETTERS

October 2, 2012

I met Beth around 2005. She is my sister-in-law. When we first met, she was working at her Subway, 24/7. We would go into the Subway and help her do things; we would go to her house and see her do a lot of things. We would all do a lot of stuff together. Beth was kind, and always going, and going with all sorts of projects.

Soon after that, Beth called and said she did not feel well. We were all shocked because we had never seen her sick, and down. It took over her body like a disease would do to a person's body. We felt helpless because we did not know what to do for her. She was always involved with her kids and so much, to where she could not do anything any longer. She was helpless. Now, she has been through so many kinds of treatments and doctors to try to get help, I feel like she has been through stages of healing. She is better some days, and not so much other days. It is a very up and down process. I don't know if I ever think she will be 100% better. She is lucky to be alive, and I know she is thankful for that, but I wish she could figure out how to get to 100%. She is still the kind, sweet, caring person that she was when we first met, but our bodies can only withstand so much. I just pray for a good outcome soon.

Mary Ked Driver

October 11, 2012

Wes and I first learned of Beth and her health issues from her Dad, when we met him in 2010. The concept of EMF poisoning was something we had never heard of at the time, which caused us to do further investigation of it and were amazed to learn how prevalent it is in our environment and even more shocking that it's never been made public knowledge about how dangerous it is for the human body. We were later able to meet Beth in April 2011, when we had the opportunity to share the health and wellness product line with her. We had learned that getting good nutrition into your body would allow the body to heal itself of many health issues and we felt this might be worth a try for her.

Beth had discovered the Water machine about a year before and wanted to share the benefits of it with us, as she had experienced an improvement in her health from drinking this water. Initially, she was not interested in trying the product, but eventually she decided to try it when she learned that it could cleanse the cells of toxins and feed the cells with good nutrition.

Through this initial association, we have become very good friends with Beth and have watched her overall health improve and rebound in an incredible way, by her shear strength, tenacity and will to live through this near-death experience. Beth is very motivated and driven to succeed, with a desire to be the best at whatever she does, including researching and utilizing numerous alternative sources for healing her body. She is an EMF survivor. Beth is still recovering, but has come a long way in improving her own health and that of her family, and has improved substantially from the first time we met her and is continuing to improve daily. Today, her mission is to spread awareness of this devastating illness, how dangerous and abundant EMF poisoning really is, and the importance of protecting your body from the exposure to it.

Beth is one of the most amazing women we've ever known and certainly is a true example of a survivor and a leader, a pioneer in this arena of EMF poisoning, which has been kept secret for so long. Through Beth's book, people are able to learn more about how EMF poisoning can happen and what one can do to improve their quality of life after severe exposure. For me, learning from Beth has enhanced my awareness of this issue and I am now able to share it with my clients, through my residential design firm, and assist them in making their homes safer through this knowledge. We whole-heartedly support Beth in her healing process and the expansion of this knowledge to everyone.

Best wishes for a full recovery,

Jenny Pippin and Wes Stearns

www.pippinhomedesigns.com

October 17, 2012

Beth,

I am very happy to hear of your recovery from Electric Magnetic Sensitivity. Though I have heard of the condition, I must admit that you are the only person I have met with the diagnosis. I am also very glad to hear that replacing your amalgams played a role in your healing. As you know, our office replaces amalgam restorations only when necessary and today we replace them with newer materials that are far more esthetic. We are always willing to replace these restorations in patients who feel that their amalgam restorations are causing health issues for them, but we cannot promise that replacing these restorations will provide patients with the same health benefits that you have experienced. For that reason I must decline any credit for your recovery. It is your persistence in getting to the bottom of your disease and not giving up on finding treatment options that helped you overcome your disease. You deserve all the credit.

Beth, it has always been a pleasure having you and your family as patients in our practice for all these many years and we look forward to helping you with your healthcare needs for many years to come. Congratulations on overcoming your disease and good luck with your book!

Dr. Jeff Phillippi

Dr. Carol E Benoit DO
Wellness and Nutrition Medicine
909 Summers Ave
Orangeburg, SC 29115

FOR BETH STURDIVANT
November 27, 2012

I first met Beth in the fall of 2010 while I was assisting Dr. Allan Lieberman in seeing patients in his Center for Occupational and Environmental Medicine in North Charleston, SC. Beth was an engaging, articulate, and very well informed patient and I found her delightful from the outset. She complained of swelling all over her body, which she believed (and still believes along with other doctors) was caused by severe electromagnetic exposure from the high tension power lines that entered her house via her yard causing chronic stress. This was my introduction to EMF illness. Beth was able to explain the connection and teach me an impressive amount about the reasons for and effects of EMF illness. She documented her claim of illness with a detailed account of her many symptoms, extending to nearly a year in which she was scarcely able to get out of bed and had considerable brain fog. By the time we met, about a year after her bedridden experience, she had learned enough through her own investigations that she was considerably improved but having a long ways to go, she remained unable to enter the Subway Shop which she owned due to all the EMF exposure there. (She still avoids going into the store.) She explained that the purpose of the attractive ceramic pendant, which she wore around her neck, was to deflect powerful, unnatural EMF radiation so that the wearer would have more protection to lessen the symptoms from this sickness. This was my introduction to EMF shields!

My friendship with Beth grew while I spent her last month together receiving treatment in Dr Lieberman's Environmental Illness center. We and several other patients spent all day together, ate together, sat in the sauna together, and got to know each other deeply. Beth then kept in touch with me while she flew to Dallas Texas, Dr. William Rea, for EMF evaluation by the most noted physician in the field. Beth subsequently sought alternative treatment at Dr. Allen's Brain Camp, which helped her more than any other treatment to date, and such was my respect for Beth's honesty and intelligence. Beth, and her remarkable parents and one of her brothers with whom I have also spent time, have become cherished lifelong friends to me, in my "inner circle" of friends for whom I will do anything that I would do for family. I continue to find in Beth and her family a wellspring of unconditional love, wisdom, experience, and a great deal of knowledge about living in health-promoting ways.

I am deeply blessed by your friendship, Beth!

Dr. Carol Benoit

November 2012

I am honored to know Beth Sturdivant and her family. From the first time I met Beth she was a fast pace worker with no time for playing around. I was 16 and she had just hired me as a worker in her store and the first thing she had me do was mop the lobby. So I proceeded to put the chairs on top of the table and she quickly let me know that's not how we do it here. Her motto was if you have time to lean you have time to clean that's why I can truly say her stores were the cleanest Subways that I have ever seen. She was a great boss who could relate to anything that you may be going thru no matter the situation and could read you before you opened your mouth and know what kind of night I had the night before lol. With this being said she would give you the shirt off her back If you needed it so that's why I had no problem working hard for her and the thing that I respect from her was she would work the long hours with you. I have never known a boss who would work open to close every day and then go home and clean and take care of her husband and kids. Beth had only one speed and that was go all the time which is the only way she knew. She inspired me in many ways letting me know you too can be a successful business owner. The number one thing that would make you successful would be your work ethic and how bad you wanted it which in return taught me that nothing was going to be given to you at any point in time in your life. She would talk to you about things that you wouldn't talk to your parents about and would also give great insight on the situation. She also inspired me to never take no for an answer or give up, you always had to look at the positive things in life no matter how it seemed. When I first noticed a change in Beth I just thought maybe she was burnt out and needed a break from her rigorous schedule. Over time you could tell that wasn't the case because she loved her stores and when I saw that every time she would come in she was in and out you knew something was going on. She never had a down look on her but you could tell that she wasn't herself never would come out and just staying around the house is not her she always was active doing something. When she started missing her son's game that's when I knew something was terribly wrong. That was never the case whatever she was doing would stop when it came time for Brandon and his sports so that was when I knew that something wasn't right. So she began to tell me that she had a sickness that I had probably never heard of and she was right. Knowing her she would explain it to me in great detail and that's when I first learned of EMF and the different life threatening things that it caused. So even though she was down, her hard work and drive for success kicked in and she began to study it and find ways to beat it and overcome the sickness that had zapped her joy and somewhat motivation for life. So for about three years she was fighting this battle and there was nothing that we could do for her, which was frustrating. Deep down inside I felt that she would beat it with her strong belief in God almighty and constant prayers from her own family. It was something that I knew very little about and could tell it was taking a toll on her by her actions. I keep you and your family in my prayers and know that you will get better and this will be something of the past that you will look back on and say that you have beaten this. WE know that the old Beth is coming back soon. Get well.

Thank you for everything and will see you over the holidays!!!
Michael Jennings

December 5, 2012

After receiving good environmental treatment, this patient got dramatically better. She was able to resume function in society. Her saga shows that an individual who was very sick can recover with environmental manipulation if the individual persists like Beth Sturdivant did. I recommend this book.

Sincerely,

William J. Rea, M.D.
Environmental Health Center – Dallas
8345 Walnut Hill Lane, Ste. 220
Dallas, TX 75231
wjr@ehcd.com

December 22, 2012

Beth's journey started on a beautiful fall day in our home peaceful and quiet with just the three of us. She was a happy content baby with lots of smiles. She has always been determined "the I can do it girl!" Beth is a caring, giving, smart, multi-tasking individual with a generous talent of entrepreneurial skills. We remember a spaghetti dinner she made by herself from scratch for the family at the young age of seven.

The long horrible quiet nightmare started in 2000 after moving in a house built too close to transmitting power lines that became more potent in 2006 as the power company added more lines. Beth's personality started changing. Month by month becoming very impatient, anxious, distracted not being able to focus, headaches, severe neck pain, severe back pain, emotional, not being able to sleep. Beth was topped her limit of stress when she endured a six month Internal Revenue Service audit for her business which put her over the cliff along with all prior problems. Beth discovered her laptop computer was causing her abdomen to hurt and swell every time she used it, resulting in an ovarian cyst. Beth spent many days traveling to physicians for help with little results due to the lack of knowledge they had on her symptoms. Beth spent days living out of a suitcase staying with different family members and friends seeking relief from retching, nausea, swelling, extreme headaches, and lack of sleep causing very dark circles under her eyes. She spent many days in the dark because light, vibrations, and noise would cause her pain and make her retch. At Beth's worst it would take her all day to enter one entry into her checkbook. The simple task of turning paper would make her retch and cause nausea. Beth's brain was in a dense fog unlike our quick thinking daughter always on her toes. Beth never gave up during this journey and was always positive encouraging others. With this dark experience she chose to rely on Lord Jesus Christ, which gave Beth a lot of hope and appreciation to know him as her rock. Each day she overcame tremendous obstacles (physical, emotional, and spiritual). Being Beth, her tenacity has allowed us to gain much valuable knowledge through her research to survive in this electrical toxic world.

As parents feeling so inadequate to know how to help and take away her pain was a big nightmare. We have been with her on this incredible journey and are so thankful to the Lord for getting our wonderful daughter back from the darkness of electrical magnetic sickness.

We love you so much!!

Mom and Dad

August, 12, 2013

Beth is true inspiration with a lot of courage! Taking her life experience and laying it on the line, with the hope of educating and helping all of us, about the hidden dangers that affect us all on this planet, that are right in your backyard. Thank you Beth.

Mike Pitts
Power Engineer
www.mikewpitts.com/safe-emf-green-power/

August 27, 2013

I am Barnabas, Beth's brother, and what I have to say is emotional for me with sadness and also happiness.

My sister has always been an upbeat person with a friendly smile and has a big heart to help her family and friends in need and also strangers. She has been an entrepreneur her whole life and has always had success in whatever she has done. Because of her work ethics and her strong will and attitude for achievement and to say "YEAH I DID THAT", "WHAT'S NEXT?"

Around 2005, I noticed some changes in my sister like mood changes, attitude on life, physical appearance and less patience for everyone. I did not understand what was going on with her and it bothered me inside but I kept it to myself. I saw my sister on several occasions after working in her front and back yard her back would go out and she would be in bed from a 1 week to 3 weeks up to a couple months. She had headaches; PMS was more painful than normal for her, her skin was starting to discolor in different areas nobody at this time could understand what was happening or why this is happening.

In 2008 my sister's life took a left turn so to speak. She was experiencing stronger headaches for longer periods of time; severe swelling in her arms, chest, stomach, neck, face, thighs, ankles, feet, and her right big toe was in constant pain. Then in 2009 she started retching (which nobody understood this type of reaction at all) for 1 to 5 minutes at a time, as well as crying uncontrollable for minutes at a time. When the weather changed from sunny skies to rain to thunderstorms, lightning, windy days, her symptoms would feel deeper and stronger inside her body and it took longer for her to have relief after the storms passed.

I noticed changes in my sister almost every day. Her mood swings of irritation of people around her she just wanted to be in the dark by herself she even said several times that she was a burden to family and friends that took the time to try to help her with discomforts of life. She had lack of energy or desire to get up out of bed due to pain to go to work and even going back to her Subway Sandwich shop from 2009 to present date. Her body would have different transformation almost daily of swelling 5 to 10 lbs depending on the different locations she was at. She drove very little due to the pain she felt with uncontrollable feelings from within her body, which it appeared to be from the invisible electrical wires small ones and large, Wi-Fi towers, power Lines, cell phones, stop lights and lighted commercial signs etc.

The shocking part to all of this experience between 2008 to 2012 was WHY this was happening to my sister.

In 2010 my sister came to live with me to see if where my location of living would help her body find relief from all her symptoms from what we have identified as EMF or A/C (Alternate Currant) waves and frequency that we can't see but is moving through our body all the time. For my sister this has severe damage internally. Her stay lasted 3 weeks but I got to witness first hand of the destruction of these silent waves and frequencies was doing to her. For example when the refrigerator would surge power, the A/C unit would turn on, washer and dryer would be running, garage door opening and closing, turning on lights or even dimming them, turning on the T.V. or radio, the ceiling fan, or baking in the oven or frying eggs on top of the stove. All had a different affect on her body inside and out of how she felt mentally, spiritually, and emotionally. Knowing that my sister has been and is still helpless from these invisible frequencies and waves and I could not comfort her by stopping her pain with the different knowledge the Lord Jesus Christ has given me was for me a very sad time in my life and will still be until we are able to find a cure for 100% recovery for her body, mind and spirit.

In closing I love my sister, Beth, and I am proud of her knowledge, wisdom, patience, kindness, tender heart, and above all her charity towards me and our family as well as friends and strangers. This is the happiness I feel inside when I get to see and hear my BEAUTIFUL SISTER BETH.

Brother Barnabas

Psalms 27:1 The Lord is my light and my Salvation; whom shall fear? The Lord is the Strength of my life; of whom shall I be afraid?

AUGUST 21, 2013

Talk about "life changing moments"... first boyfriend, first kiss, graduation, wedding day, birth of children, the day the doctor tells you your son has Aspergers. It was like the world stopped and I couldn't breathe. I was shocked, knew there had to be a mistake. I went to school to become a teacher; I was trained for working with children with special needs. I spent the next few years going to top specialists in New York to figure out how can I help my child be ok. Why did this happen? Of course my son was going to be ok, but the doctors gave me no answers. Just large out of pocket medical bills. It was April 2011 another "life changing moment"... I met Beth. We began talking about my son and I told her the story. Beth looked at me and said "so have you healed him yet?" I thought to myself what is this person saying? With my education and all the doctors no one has ever said this to me. That was the beginning of my journey to heal my son. Beth taught me words like "blood brain barrier" and "glutathione", words I never heard of before. She introduced me to alkaline water, essential oils, amino acids, photon genius, playing games with a purpose, etc. One story I would like to share is as simple as a haircut. My son always fought and cried when it was time for his haircut. It was always a very stressful experience and I mean stressful. Beth explained to me that it was the buzzer, which was very painful for him. From that day on we used old fashion scissors and it wasn't a problem anymore. Clearly, Beth wasn't in good health herself, but she never once complained. She was always there to help people and was always busy working. Moving forward with a positive attitude healing herself while helping others. I am so thankful to have met Beth; she has been an inspiration for me. Teaching me important information that I didn't learn in school and training me with techniques that will help my son. One thing I pray for is a full recovery for Beth. I hope this book will reach and help as many people as possible.

Thank you, Beth

DEBORAH CLESE
NY/NC State Certified EC Teacher

September 7, 2013

As an editor of this book and a close friend with Beth, I can attest to the validity of her story. I've known Beth since she was nine years old. The first time I met her was on the school bus and her mother was our driver. I grew to know her family, their values, practices, and beliefs. Beth's parents were adamant in teaching their children about taking care of their bodies and health as well as learning about God. I have never met a couple more disciplined in their beliefs and practices than Beth's parents. They worked hard and did the best they could raising five children. Beth came from extremely humble beginnings and portrayed strong, moral, and ethical work characteristics. I knew Beth before and during her health crisis. She embodies a quiet tenacious spirit with a "can-do attitude". She strives to be positive and encouraging. Beth loves her husband and children very deeply. She is grateful to her parents and siblings, always wanting the best for every one she knows or meets. Beth works at helping herself and others to get beyond the pain or suffering from life's trials. She teaches those that will listen that through prayer, dependence, and reliance on God you can survive as well as thrive. Beth has suffered immensely and can help others if they choose to listen while being open minded to what she has to share from her journey. She's a great and gracious friend to all those that know her. I'm grateful that our paths have crossed on this side of heaven. She truly is "One N A Million"!

By - Caroline D. Green
carolinedgreen.com

September 22, 2013

I will never forget the day I first met Beth. This lady came in to my home, placed a wire that looked like an extension cord around her ankle, and plugged it into the grounding part of my electrical outlet. That's different, I thought! We had initially met each other over the telephone and we both had gone to the same brain center at different times for treatment. Beth was searching for help from EMF poisoning and I had gone for help for neurotoxin poisoning. Our friendship and journey to get well began that day she came over to my house. Eventually, in her quest to help, Beth found some whole food products that provided the nutrition our bodies needed in order to increase the performance of our immune system's ability to fight our sicknesses as well as improve our overall health.

Beth has been such an inspiration to me as I have seen her never stopping for answers, never giving up, and fighting for her life! We both have a passion to help others never experience the journey we had to travel. This book is the map! It can help people become aware of EMF conditions and how to protect themselves. You will discover through reading her book the many hidden obstacles out in our world today. Beth's book will help people everywhere to stay on course to their final destination for good health! Thank you Beth for never quitting! You helped save my life and I thank God for you and your book!

Denise Blankinship, Airline Captain

This book is essential reading to understand how harmful, if not deadly, exposure to electronic and magnetic frequencies are. This accurate effort should be read by all.

Russell M. Bianchi
Managing Director of Adept Solutions, Inc.
A Global Product Development Firm

*Backyard Secret Exposed* is an eye opening book of one woman's journey through being poisoned by what we all take for granted daily, electricity. When I was introduced to Beth and saw for myself the effects on her body as well as the agony to her family that EMF poisoning had done, I was completely shocked! To see and learn of her symptoms from the exposure to high levels of electromagnetic fields was hard to digest when knowing that our society has exemplified those symptoms in various forms but are prescribe drugs as the way to treat the issues. Furthermore, instead of educating and teaching people on how to protect themselves is simply unforgivable. I applaud Beth and her efforts to educate us all on how to protect ourselves and families so we can enjoy modern world technologies while taking care of our health in the process. Her road has been tough. However, I commend her efforts in recovery and for NOT giving up, nor accepting the status quo answers that all the educated doctors were giving her. I will be standing up with her to bring about EMF awareness as I further work to prevent myself and family from being harmed in the future.

Mona Lisa Wilcox
Florida State Certified General Contractor
Health and Wellness Advocate

8-10-2013

I met Beth Sturdivant over year ago, and my life has not been the same since. She helped me get out of the house after seven years from being sick. She is the most fascinating person I know. I am blessed that she is my friend. I admire her courage and determination so much... I have enjoyed working with Beth as she is so positive, uplifting, and never gives up.

Everyone should read her book to learn more of what to become aware of in this 21st century gadget world.

Marian Foster
Retired Italian Restaurant owner Long Island, New York

September 18, 2013

I am very honored to not only be asked to write a testimony for this amazing book but I am very proud to be the daughter of this incredible woman who wrote the book! My name is Brandi Sturdivant and Beth Sturdivant is my mother. Our lives changed dramatically as a family in 2009 when my mother's health noticeably began to fail and none of us had any idea why!

In order to give you a little background bio on my mom, she is basically a Wonder Woman! She was always a very active part of my life. She never sat still! She always had a project going whether it was scrapbooking, remodeling rental homes, or taking my brother and me to where we needed to go as her children. You get the idea? She has been very influential in many people's lives, not just mine, but she has many employees that worked for her as teenagers that are still in contact with her. I noticed a difference in my mom the last semester that I was in high school, which was in 2007. She just seemed more irritable such as getting stressed out by everything when normally she handled everything so smoothly before like it was no big deal. My brother and I would say to each other, "Something's going on with Mommy," but we had no idea what or what to do.

After a while she started having other things happening, but we didn't realize they were symptoms of something much bigger and more intense things to come. She started having dizzy spells, extremely intense headaches, and tremendous anxiety. Her back would hurt and went out for extended periods of time. In 2009, we started noticing she was swelling and thought she's experiencing pregnancy symptoms. Her stomach swelled up, her breasts were leaking some form of liquid and her cheeks turned red while rashes appeared on her chest and stomach areas. My mom had come to visit me at college and she said "Brandi look at my stomach its swelling out". At that point I started to become concerned because I was thinking if she's not pregnant then something's really wrong. The next time I came home, we went to the emergency room and she took a pregnancy test. The test came back negative and the doctors couldn't explain her symptoms.

After figuring out what was wrong with her, we all had to make DRASTIC lifestyle changes. Many of which everybody should adopt to be preventative. As examples, WIFI in the house had to be turned off, we stopped using the microwave, no cell phones could be charged in a room where my mom was, and many times we couldn't have cell phones around her until we got some of those Diodes she found that worked for us.

All I can say is this is the most life changing experience I have had to endure. I have learned so many things, which will forever change my life. I am thankful that my mother is a survivor and that she never gave up during this challenging time in her life. I am so grateful and happy that she took what was and is a negative experience and turned it into something positive. I am joyful that this journey we took together with my mother's experiences has opened our eyes and hearts to help others through an opportunity I have by starting a foundation called The Academy for Environmental Sickness. This foundation is designed to help educate people for preventative measures as well as assist others who are chronically ill through providing hope and natural products that could potentially turn their lives around.

I Love you Mom! You're the best woman I know and I am glad we have many more years together!

Brandi Sturdivant

September 22, 2013

I've had months to write this testimonial for my mom's book but haven't budged to do so until crunch time. Her book is heading toward publication now. I'm very privileged to do so considering the caliber of a lady she is in all life aspects.

When I was about 14 years old I noticed a beyond drastic change in my mom. It all began when she started letting me drive her around before I even had my learner's permit because she didn't feel well. Being young, I thought it was cool at the time because I enjoyed breaking the rules and getting away with it having her on my side while I was doing it. HA- HA! I knew my mom wasn't feeling well, but I just didn't pay it any attention because I didn't want to consider what could happen to her. Eventually driving her around turned into driving alone even before I had my permit. She just didn't feel well enough to do anything but lay in her bed. Mom gave me money and I would go to the grocery store doing the shopping buying what I needed for the week. She also allowed me to drive up the street to visit friends since they were close to the house because she was too sick to drive me herself. This was all fine and dandy to me as I was having fun.

After a while the fun wore off and when my mom had to move out of our house in order to go stay at my grandparents it hit me just how sick she really felt. I remember walking into the room and seeing her lying on a bed with tin foil all around her while wearing sunglasses over her eyes in the pitch dark. I honestly thought I was looking at my mom on her death bed. For a teenager who was used to seeing their mother on the go all the time to seeing her lying on the bed in this condition was over-whelming on top of the fact no one had a clue what was wrong or what to do for her. I didn't stay long visiting that day because I had to spend some time alone accepting the fact that something wasn't right with mom and I had to consider what may be ahead of me.

That was a hard thing to go through no matter what age a person is, but being a teenager who was used to having a great mother around doing a lot that was difficult to come to grips with her possible passing. Over the course of her sickness, when I was with my mom I couldn't have anything that involved electricity. No cell phone, watching TV, or really anything because it would cause her to retch and I didn't like seeing that one bit. Although all of this was the case, I still had to rip and run alone handling what I needed to on my end in order to make sure everything was smooth with school and sports. However, through it all, my mother still found a way to take care of me, my sister, and dad no matter how sick she felt. I know she isn't all the way back to 100%, but I feel like I can see the light at the end of the tunnel for her recovery to wellness.

I just want you to know, Mom, that I am very proud of you and all of your accomplishments. You are the greatest woman that I know and I appreciate everything that you've done for this family. You were the one that held it down for us all, even when you were at your lowest point. I salute you for that and will forever love you for it. Congratulations, Author Beth Sturdivant, you deserve it!

Brandon Sturdivant

September 11, 2013

I was introduced to Beth through my sister. We actually met over the phone while Beth was in Idaho at a brain camp session. We discovered that we both lived within driving distance of each other in the state of South Carolina. Her journey for wellness has greatly impacted me and my family, as well as my own journey for the same quest. We have become friends in this process of getting our health back.

Donnie & Cheri Spires - Owners of Palmetto Home Style Bath & Kitchen

www.palmettohomestyle.com

# ACKNOWLEDGEMENTS

Acknowledgement will never show my sincere thankfulness, only reference.

* To my Lord and Savior first and thank Him for giving me the ability to solve this problem.
John 3:16

* I want to thank my awe-inspiring children who endured this race and finished strong with me. Brandi: your determination to help ease my pain and researching remedies. Brandi: also finishing college not giving up. Brandon: your dedication to me always asking, "What can I do for you, Mommy?" Brandon: working hard, preparing yourself with Goals. Doing what only you could set out to do after knee surgery. Congratulations for walking on USC Gamecocks Football Roster and not giving up. You both are the highlight of my life and I cannot begin to acknowledge all your support. Just know I love you and cherish you every day of my life, my true gifts from God. Love you, Mommy

* My wondrous parents who gave unconditional support, help and time. Mom, I know all the helpless days you spent with me wondering what you could do for me, the entire doctor trips out of town. Dad just allowing Mom to be gone for weeks at a time with me and your prayers for my family and me. Thank you both with all my heart. I love you, Beth

* Johnny, my husband, for hanging in here and learning to cope with the sickness, who loves me, and whom I so appreciate for running the restaurant every day and keeping the family afloat. I pray we can get this behind us and live the life we well deserve, happily and healthy. I love you very much and appreciate you, Beth

* Ben, Barak, Barnabas, and Bekah, my siblings, who helped me with travels, living arrangements, and trying to get me through the moment. I love you all, Beth

* Extended family who came to visit, pray, and help in any way you knew how. Love always, Beth

* Many doctors who gave me one answer at a time.

* I want to thank my Great friend of 33 years Caroline Green. I must say you definitely have gotten the bird's eye view of this journey by witnessing me staying in your home and editing this book. Thank you for being so conscientious of my sensitivity at all times but mostly using what you have learned and teaching Ashley. I do not believe anyone could follow my story like you and keep my voice. You have a real talent. I wish you much success with your personal book Decrees, Declares, & Prayers. I love and appreciate you. Your Best Friend, Beth www.carolinedgreen.com

* I want to thank my great friend and mentor, Omega Autry. Like we say, the parallel lives we have lived and experiences together have been incredible, along with the memories I will always cherish. Thank you for proof reading and always giving a helping

hand. I am so blessed to call you my friend. I appreciate and love you, Beth

* Thank you to my new friends that have joined my path to healing and bringing awareness to the EMF sickness: Dr. Carol Benoit, Denise Blankinship, Deborah Clese, Julie Epple, Marian Foster, Mike Pitts, Cheri Spires, Mona Lisa Wilcox, and Nanette Young.

* Julie Epple, thank you for proof reading my book at the beginning. www.nojunkjulies.com

* I would like to thank Wayne Cooper for taking his time and treating my book as though it was his. Wayne, you are a Godsent with a lot of knowledge and kindness. May the Lord continue to bless you and your family always. Prayers and Blessings, Beth

* Tracey Coon you really have a gift within. I am so grateful for your extraordinary talent. I want to thank you for making my vision be seen through my book cover and logos. May the Lord bless you and your business. Thank you, Beth

* Thank you to my friend, Chris Smith, for giving me more freedom than anyone. I cannot thank you enough for the grounding gadget you perfected that has allowed me to at least be able to deal with public places when it's been a must. I would never been able to see my son play in a gym without it. I am very honored to have had you as a friend for the past 35 yrs. I will always cherish our friendship, Beth

* Nick, my attorney, who helped land the landmark case win for disability when all odds were stacked against me.

* My EMF community I know and will get to know who are suffering with no answers, feeling hopeless and helpless. I am very blessed to have all this support. I could not have accomplished this by myself. Love to all and pray one memory at a time.

Blessings and Prayers, Proverbs 3:5-6 "Trust in the Lord with all thine heart; and lean not unto thine own understanding. In all thy ways acknowledge him, and he shall direct thy paths".

**zija** | INDEPENDENT DISTRIBUTOR

www.onenamillion.myzija.com

## BETH STURDIVANT
## 704.458.1010

To Help Others
Send all Donations to:
**The Academy For Environmental Sickness**
**PO Box 455, Pineville, NC 28134**
**onenamillion@me.com**

Made in the USA
Charleston, SC
25 November 2013